THE COMPLETE GUIDE TO BACKYARD DAIRY FARMING: FROM NOVICE TO EXPERT

A Practical Handbook for Successful and Rewarding Dairy Farming at Home

James B. Mike

Bonus: The processes of making butter, cheese, yogurt, and other delicious and nutritious dairy products.

Table of Contents

Introduction
1. The Importance of Backyard Dairy Farming

Chapter 1
Getting Started
2. Assessing Your Resources and Setting Goals
2.1 Evaluating Available Space and Resources
2.2 Defining Your Objectives and Vision
2.3 Planning for Success: Budgeting and Time Management

Chapter 2
Choosing Dairy Animals
3. Understanding Different Dairy Breeds
   3.1 Cow Breeds: Characteristics and Milk Production
   3.2 Goat Breeds: Suitable Options for Backyard Farming
   3.3 Sheep Breeds: Considerations for Small-Scale Dairy Farming

Chapter 3
Housing and Facilities
4. Designing a Functional and Comfortable Shelter
4.1 Housing Requirements for Dairy Animals
4.2 Constructing a Milking Parlor and Handling Facility
4.3 Waste Management and Environmental Considerations

Chapter 4
Feeding and Nutrition
5. Nutritional Needs of Dairy Animals
5.1 Understanding Basic Nutrient Requirements
5.2 Forage Management and Pasture Grazing
5.3 Supplementing Feed and Ensuring a Balanced Diet

Chapter 5
Milking and Milk Handling
6. Milking Techniques and Equipment
6.1 Proper Milking Practices for Quality Milk Production

6.2 Selecting and Maintaining Milking Equipment
6.3 Handling, Cooling, and Storing Milk Safely

Chapter 6
Health and Disease Management
7. Maintaining Optimal Health and Preventing Diseases
7.1 Vaccination Protocols and Parasite Control
7.2 Identifying and Treating Common Health Issues
7.3 Biosecurity Measures and Disease Prevention Strategies

Chapter 7
Dairy Product Processing
8. Exploring Value-Added Dairy Products
8.1 Butter, Cheese, Yogurt, and Other Products
8.2 Safe Processing Techniques and Equipment
8.3 Marketing and Selling Your Dairy Products

Chapter 8
Troubleshooting and Problem-Solving

9. Overcoming Challenges in Backyard Dairy Farming
9.1 Managing Seasonal Variations and Weather Conditions
9.2 Reproduction and Breeding Challenges
9.3 Addressing Milking and Production Issues

Chapter 9
Scaling Up and Expanding
10. *Taking Your Backyard Dairy Farming to the Next Level*

Introduction

1. The Importance of Backyard Dairy Farming

Let's dive into the benefits, joys, and economic potential of backyard dairy farming.

The Benefits of Backyard Dairy Farming

Backyard dairy farming offers a wealth of advantages beyond just the fresh, high-quality dairy products it can provide. One of the primary benefits is the ability to have control over the entire process, from the care and feeding of the animals to the production and processing of the dairy goods. This level of involvement can be incredibly rewarding, as you can ensure your animals are treated humanely and that your dairy products are free from any unwanted additives or preservatives.

Another significant benefit is the potential for cost savings. By producing your own dairy products, you can avoid the markups and transportation costs associated with store-bought items. This can translate to significant savings over time, making backyard dairy farming a financially prudent choice, especially for families or individuals looking to reduce their grocery bills.

Moreover, backyard dairy farming can be an excellent way to teach children about the origins of their food and the importance of sustainable agriculture. Involving them in the daily tasks of caring for the animals and processing the dairy can instill a sense of appreciation and responsibility that will serve them well throughout their lives.

The Joy of Producing Your Own Dairy Products

One of the most rewarding aspects of backyard dairy farming is the joy of producing your own dairy products. There's something incredibly satisfying about watching a calf or goat kid grow, and then being able to turn their milk into a variety of delicious and nourishing foods, such as cheese, yogurt, butter, and ice cream.

The process of cheesemaking, for example, can be a true work of art. Experimenting with different recipes, aging techniques, and flavor profiles can be both challenging and immensely gratifying. Sharing your homemade cheeses with friends and family can be a wonderful way to connect and bond over a shared appreciation for the fruits of your labor.

Similarly, the process of churning your own butter or making your own yogurt can be incredibly fulfilling. Knowing that you've created these staple dairy products from the milk of your own animals can instill a deep sense of pride and accomplishment.

The Economic Potential of Backyard Dairy Farming

While backyard dairy farming may not be a path to vast riches, it can offer significant economic benefits. By reducing your household's reliance on store-bought dairy products, you can save a considerable amount of money over time. Additionally, if you have a surplus of dairy products, you may be able to sell or barter them within your local community, generating a small additional income stream.

Some enterprising backyard dairy farmers even take their operations to the next level, selling their wares at farmers' markets or through community-supported agriculture (CSA) programs. This can be a great way to connect with like-minded individuals and share the fruits of your labor with a wider audience.Kcr8jzk

It's important to note that the economic potential of backyard dairy farming can vary widely depending on the size of your operation, the laws and regulations in your area, and the demand for your products within your local community. However, even a small-scale backyard dairy can provide a valuable source of income and self-sufficiency.

The Importance of Backyard Dairy Farming

In a world increasingly dominated by industrialized agriculture and mass-produced food, the importance of backyard dairy farming cannot be overstated. By taking control of your own dairy production, you're not only ensuring the quality and provenance of your food, but you're also contributing to a more sustainable and resilient food system.

Backyard dairy farming can serve as a powerful counterpoint to the environmental and ethical concerns surrounding large-scale industrial dairy operations. By reducing your reliance on commercially produced dairy products, you're lowering your carbon footprint and supporting a more humane and ecologically-responsible approach to food production.

Moreover, the act of backyard dairy farming can be a transformative experience, reconnecting you with the rhythms of nature and the satisfaction of providing for yourself and your family. It's a way to cultivate a deeper appreciation for the origins of your food and the hard work that goes into its production.

In a world that can sometimes feel increasingly disconnected and impersonal, backyard dairy farming offers a chance to rediscover the joy and fulfillment that can come from a more hands-on, self-sufficient approach to living. It's a powerful reminder that we all have the capacity to take greater control over our own sustenance and well-being.

Chapter 1

Getting Started

2. Assessing Your Resources and Setting Goals

2.1 Evaluating Available Space and Resources
Before you dive headfirst into the world of backyard dairy farming, it's crucial to take a step back and carefully evaluate the resources and space you have available. This assessment will help you set realistic goals and ensure that your dairy farming endeavor is a sustainable and successful one.

Assessing Your Available Space
The amount of land and facilities you have will largely determine the scale and scope of your backyard dairy operation. Start by taking a close look at your property and *asking yourself a few key questions:*

- *How much total land area do you have available?* This will dictate the number of dairy animals you can comfortably accommodate.

- *Do you have a dedicated area for a barn, paddock, or pasture?* Dairy cows, goats, or other dairy animals will need designated spaces for housing, grazing, and exercise.

- *Is there access to clean water sources,* both for your animals and for cleaning/sanitizing equipment?

- *Are there any zoning restrictions or homeowner's association rules* that limit the types of animals or farming activities allowed?

Carefully evaluating the physical space you have to work with will help you select the right dairy animals and plan your setup accordingly. *For example*, if you only have a small backyard, a couple of dairy goats may be a more feasible option than a full-size dairy cow.

Assessing Your Resources

In addition to physical space, you'll also need to take stock of your other available resources, such as:

Financial Resources:

- How much can you reasonably afford to invest in startup costs, such as purchasing animals, building/renovating facilities, and acquiring necessary equipment?

- Do you have access to financing options like loans or grants to supplement your personal funds?

- Can you absorb the ongoing operational expenses, including feed, veterinary care, utilities, and labor? Resources:

- Do you have the time, energy, and physical capability to handle the daily tasks of dairy farming?

- Can you rely on family members or hired help to assist with the workload?

- Do you have access to educational resources, such as local extension services or experienced dairy farmers who can provide guidance?

Skill and Knowledge Resources:
- What is your current level of experience and expertise in animal husbandry, dairy processing, and other relevant areas?

- Are you willing and able to invest in training, workshops, or mentorship opportunities to build your skills?

- Do you have a reliable source of information, such as this comprehensive guide, to help you navigate the ins and outs of backyard dairy farming?

Setting Realistic Goals
With a clear understanding of your available space and resources, you can now start setting

goals for your backyard dairy operation. These goals should be specific, measurable, achievable, relevant, and time-bound (SMART). ***Some examples of goals you might consider include:***

- Establish a small herd of 2-3 dairy goats and begin producing milk for personal consumption within the first year.

- Construct a 500-square-foot barn and fenced paddock to accommodate a single dairy cow within the next 18 months.

- Learn how to properly milk, process, and preserve dairy products, with the goal of selling surplus milk, cheese, and yogurt at a local farmers' market within 2 years.

- Reach a sustainable level of dairy production that can provide at least 50% of your household's dairy needs within 3 years.

Remember, your goals should be realistic and aligned with the resources you have available.

It's better to start small and gradually scale up your backyard dairy operation than to overcommit and risk running into challenges you're not equipped to handle.

By carefully evaluating your space and resources and setting achievable goals, you'll lay the foundation for a rewarding and sustainable backyard dairy farming experience.

2.2 Defining Your Objectives and Vision

Before embarking on your backyard dairy farming journey, it's essential to clearly define your objectives and vision. This will serve as a guiding light throughout the process and help you make informed decisions.

First, ask yourself: Why do you want to start a backyard dairy farm? Is it to provide fresh, nutrient-rich milk for your family? To become more self-sufficient? To earn additional income? Or perhaps a combination of these? Identifying your primary motivations will help you shape your overall vision and goals.

Next, consider the scale and scope of your backyard dairy farm. Do you envision a small setup with just a couple of dairy animals, or do you have plans to expand and possibly sell your surplus milk and dairy products? Answering these questions will determine the resources you'll need to acquire and the level of commitment required.

It's also important to think about the lifestyle changes that come with owning a backyard dairy farm. Dairy animals require daily care and attention, so be prepared to adjust your daily routine to accommodate the needs of your herd. Determine whether you have the time, energy, and physical capability to handle the hands-on tasks involved in dairy farming.

Assessing Your Resources

Once you've defined your objectives and vision, it's time to assess the resources you have available. This includes both tangible and intangible resources.

Tangible Resources:
- *Land:* Determine the size of your available land and whether it can accommodate the number of dairy animals you envision. Consider factors like grazing areas, shelter, and access to clean water.
- *Facilities:* Assess the existing structures on your property, such as barns, sheds, or outbuildings, and evaluate their suitability for housing your dairy animals and processing milk.
- *Equipment:* Make a list of the necessary equipment, such as milking machines, milk storage tanks, and processing tools, and determine what you already have or need to acquire.
- *Finances:* Estimate the startup costs, including the purchase of dairy animals, feed, veterinary care, and any necessary infrastructure upgrades. Ensure you have the financial resources to sustain your backyard dairy farm.

Intangible Resources:

- ***Knowledge and Skills:*** Evaluate your existing knowledge and skills in dairy farming. If you're a beginner, consider seeking out educational resources, workshops, or mentorship opportunities to build the necessary expertise.

- ***Time and Labor:*** Assess the time and labor you can personally commit to the daily operations of your backyard dairy farm. Determine whether you'll need additional help, such as family members or hired labor, to manage the workload.

- ***Support Network:*** Identify local resources, such as cooperative extensions, dairy associations, or experienced backyard farmers, who can provide guidance and support throughout your journey.

Setting Goals

With your objectives, vision, and resources clearly defined, it's time to set specific, measurable, achievable, relevant, and time-bound (SMART) goals. These goals will serve as milestones to work towards and help you track your progress.

Here are some examples of SMART goals for a backyard dairy farm:
- "Acquire two Jersey cows and construct a suitable barn by the end of the year."
- "By the end of the first year, produce and process 5 gallons of raw milk per day for personal use."
- "Set up a small cheese-making business and sell 10 pounds of artisanal cheese each month by the end of the second year."
- "Implement a rotational grazing system and achieve a 20% reduction in feed costs within the first 18 months."

Remember, these goals should align with your overall vision and be tailored to your specific resources and constraints. Review and adjust your goals regularly as your backyard dairy farm evolves.

By clearly defining your objectives and vision, assessing your resources, and setting SMART goals, you'll lay a solid foundation for your backyard dairy farming journey. This approach will help you make informed decisions, allocate your resources effectively, and work towards the successful realization of your dairy farming dreams.

2.3 Planning for Success: Budgeting and Time Management

Embarking on a backyard dairy farming venture can be an incredibly rewarding experience, but it also requires meticulous planning and organization to ensure long-term success. In this comprehensive guide, we'll dive deep into the essential components of budgeting and time management - two crucial pillars that will set you up for triumph in your backyard dairy farming endeavors.

Budgeting for Backyard Dairy Farming

Crafting an accurate and realistic budget is the foundation upon which your backyard dairy farming operation will stand. It's the roadmap that will guide you through the financial landscape, helping you make informed decisions and avoid costly pitfalls.

Identifying Startup Costs

The first step in building your budget is to identify all the startup costs associated with establishing your backyard dairy farm. ***This includes, but is not limited to:***

- Land acquisition or lease fees
- Fencing and shelter construction
- Dairy cattle purchase or leasing
- Feed and water systems
- Milking equipment (e.g., buckets, pails, milking machine)
- Refrigeration and storage for milk and dairy products
- Licensing and regulatory compliance fees

By carefully accounting for each of these expenses, you can create a comprehensive snapshot of the initial financial investment required to get your backyard dairy farm up and running.

Calculating Ongoing Operational Costs

Once you've tackled the startup costs, it's time to turn your attention to the ongoing operational expenses. These include:

- Feed and forage costs
- Veterinary care and medication
- Utilities (electricity, water, heating/cooling)
- Maintenance and repairs for equipment and infrastructure
- Labor costs (if hiring help)
- Transportation and fuel
- Supplies (e.g., cleaning products, packaging materials)
- Marketing and sales expenses

Meticulously tracking these recurring costs will help you understand the monthly, quarterly, and annual financial obligations of maintaining your backyard dairy farm.

Projecting Revenue and Profit

With a clear understanding of your startup and operational costs, you can now turn your attention to projecting your potential revenue and profit. *This involves estimating:*

- Milk production and sales
- Cheese, butter, or other dairy product sales
- Income from selling excess livestock or breeding stock
- Any additional revenue streams, such as agritourism or educational tours

By carefully balancing your expenses against your anticipated income, you can develop a comprehensive budget that will guide your decision-making and help you achieve your financial goals.

Mastering Time Management in Backyard Dairy Farming

Effective time management is crucial in the world of backyard dairy farming, where the demands on your time can be overwhelming. By implementing strategic time management techniques, you can maximize your productivity, reduce stress, and ensure the smooth operation of your dairy farm.

Establishing a Daily Routine

One of the most critical aspects of time management in backyard dairy farming is the establishment of a consistent daily routine. This routine should incorporate all the necessary tasks, *such as:*

- Milking the cows
- Feeding and watering the livestock
- Cleaning and maintaining the dairy facilities
- Monitoring the health and well-being of the animals
- Processing and storing the dairy products

By creating a structured schedule and sticking to it, you can ensure that all essential tasks are completed efficiently and on time.

Prioritizing and Delegating Tasks

As your backyard dairy farm grows, you may find yourself overwhelmed with the sheer volume of tasks that need to be completed. In such cases, it's essential to prioritize your responsibilities and delegate certain tasks to others, if possible. ***This could involve:***

- Identifying the most time-sensitive and critical tasks
- Delegating less-critical tasks to family members or hired help
- Outsourcing specific functions, such as bookkeeping or marketing
- Utilizing technology and automation to streamline processes

By carefully prioritizing and delegating, you can free up valuable time to focus on the most essential aspects of your backyard dairy farming operation.

Embracing Efficiency and Productivity
In addition to establishing a routine and delegating tasks, there are various other

strategies you can employ to enhance your efficiency and productivity as a backyard dairy farmer. ***These include:***

- Leveraging checklists and task management tools
- Batch-processing similar tasks to minimize time spent transitioning
- Scheduling regular breaks and rest periods to prevent burnout
- Continuously evaluating and improving your time management practices

By implementing these strategies, you can ensure that your backyard dairy farming endeavors are not only financially stable but also sustainable in the long run.

Exercises and Practical Steps

To put the principles of budgeting and time management into practice, consider the following exercises and practical steps:

1. *Create a Detailed Startup Cost Worksheet*: Divide your startup costs into categories (e.g., land, livestock, equipment) and estimate the expenses for each. This will help you understand the initial financial investment required.

2. *Develop an Operational Cost Tracker*: Create a spreadsheet or use a budgeting app to track your monthly, quarterly, and annual operational expenses. Review this regularly to identify areas for cost-saving.

3. *Forecast Your Revenue Streams:* Estimate your potential milk, dairy product, and livestock sales based on your herd size, production rates, and market prices. Adjust your projections as you gain experience.

4. *Establish a Daily Dairy Farm Routine:* Write out a detailed schedule of your daily tasks, including milking times, feeding, cleaning, and processing. Stick to this routine to maximize efficiency.

5. Identify Time-Saving Opportunities: Evaluate your current processes and look for ways to streamline or automate certain tasks, such as using automated feeders or milking systems.

6. Delegate and Outsource: Consider hiring part-time help or outsourcing specific functions, such as bookkeeping or marketing, to free up your time for more essential tasks.

7. Utilize Productivity Tools: Experiment with various checklists, task management apps, and other digital tools to help you stay organized and on track.

By consistently applying these exercises and practical steps, you'll be well on your way to building a successful and sustainable backyard dairy farming operation.

In conclusion, effective budgeting and time management are essential for achieving long-term success in backyard dairy farming. By carefully planning your startup and operational costs, projecting your revenue, and implementing efficient time management strategies, you can lay the groundwork for a thriving and profitable backyard dairy farm. Remember, the key to success lies in your ability to balance the financial and operational aspects of your venture, allowing you to focus on the rewarding aspects of dairy farming and enjoy the fruits of your labor.

Chapter 2

Choosing Dairy Animals

3. Understanding Different Dairy Breeds

3.1 Cow Breeds: Characteristics and Milk Production

Choosing the right dairy cow breed is a critical decision for any aspiring backyard dairy farmer. Each breed offers unique characteristics, strengths, and milk production capabilities that must be carefully considered to ensure the success of your dairy operation. In this guide, we'll explore the most popular dairy cow breeds, their distinctive traits, and how to select the best fit for your backyard setup.

Understanding Dairy Cow Breeds

Dairy cows can be broadly categorized into two main groups: *European breeds* and *American breeds*. European breeds, such as Holsteins, Jerseys, and Guernseys, are renowned for their high milk production, while American breeds, like Milking Shorthorns and Ayrshires, are known for their adaptability and hardiness.

European Breeds

Holsteins: The most widely recognized dairy breed, Holsteins are renowned for their exceptional milk production. They are typically large, black-and-white in color, and can produce up to 25 liters of milk per day. Holsteins are versatile and thrive in a variety of climates, making them a popular choice for many dairy farmers.

Jerseys: Smaller in size compared to Holsteins, Jerseys are known for their rich, creamy milk. They are often fawn-colored and can produce up to 15 liters of milk per day. Jerseys are known for their gentle temperament and efficient feed-to-milk conversion ratio.

Guernseys: Originating from the island of Guernsey, this breed is characterized by its distinctive golden-colored coat and high-quality milk. Guernseys can produce up to 20 liters of milk per day and are known for their hardiness and longevity.

American Breeds

Milking Shorthorns: A versatile breed that excels in both milk and beef production, Milking Shorthorns are known for their adaptability and calm temperament. They can produce up to 15 liters of milk per day and are well-suited for smaller backyard operations.

Ayrshires: Originating from Scotland, Ayrshires are known for their high-protein and high-fat milk. They can produce up to 18 liters of milk per day and are well-suited for grazing-based systems.

Factors to Consider When Choosing a Dairy Breed

When selecting a dairy cow breed for your backyard operation, consider the following factors:

1. Milk Production: Assess your daily milk needs and choose a breed that can meet or exceed your requirements.

2. Herd Size: Consider the space and resources available for your herd. Smaller breeds like Jerseys may be more suitable for smaller backyards.

3. Climate Adaptability: Evaluate your local climate and choose a breed that is well-suited to thrive in your region.

4. Temperament: Opt for a breed with a calm and docile temperament, especially if you have limited experience handling livestock.

5. Feed Efficiency: Consider the breed's ability to convert feed into milk, as this can impact your overall operating costs.

Practical Steps for Choosing the Right Dairy Cow

1. Research and Compare Breeds: Thoroughly research the different dairy cow breeds, their characteristics, and milk production capabilities. Use online resources, breed associations, and seek advice from experienced dairy farmers.

2. *Visit Local Farms:* Arrange visits to local dairy farms to observe the cows in person. This will give you a better understanding of their size, temperament, and overall suitability for your backyard setup.

3. *Consult with a Veterinarian:* Seek the guidance of a veterinarian who specializes in dairy cattle. They can help you assess your specific needs and recommend the most suitable breed for your backyard dairy operation.

4. *Start Small:* If you're a beginner, consider starting with a single cow or a small herd. This will allow you to gain hands-on experience and gradually scale up your operation as you become more proficient.

5. *Develop a Feeding and Care Plan:* Establish a comprehensive feeding and care plan that aligns with the specific needs of your chosen dairy cow breed. This will ensure the health and productivity of your herd.

By understanding the characteristics and milk production capabilities of different dairy cow breeds, you can make an informed decision and choose the perfect fit for your backyard dairy farming endeavor. Remember, the success of your backyard dairy operation largely depends on selecting the right dairy cow breed that meets your unique needs and expectations.

3.2 Goat Breeds: Suitable Options for Backyard Farming

Raising dairy goats can be an incredibly rewarding experience for those interested in backyard farming. Goats are versatile, hardy animals that can thrive in a variety of settings, from small urban lots to larger rural properties. When it comes to selecting the right goat breed for your backyard dairy farm, *there are several excellent options to consider.*

Nubian Goats

Nubian goats are a popular choice for backyard dairy farmers due to their high butterfat content and rich, creamy milk. These large-framed goats are known for their distinctive long, pendulous ears and roman nose profile. Nubians are generally gentle in temperament and can produce between 1-2 gallons of milk per day at their peak. They tend to do well in warmer climates and can be excellent browsers, keeping weeds and vegetation in check.

Alpine Goats

Alpine goats are a versatile Swiss breed that excel in both milk production and meat. They are medium-sized with upright ears and a straight facial profile. Alpines are hardy, adaptable animals that can thrive in a variety of environments. On average, an Alpine doe can produce 1-2 gallons of milk per day. Their milk has a moderate butterfat content, making it well-suited for cheesemaking and other dairy products.

Lamancha Goats

Lamancha goats are known for their distinctive small, elf-like ears. These medium-sized dairy goats are renowned for their high milk production, with does capable of yielding up to 2 gallons per day at their peak. Lamanchas are calm, intelligent animals that are well-suited for both small and large backyard farms. Their milk has a rich, creamy texture and is excellent for cheese, yogurt, and other dairy products.

Pygmy Goats

For those with limited space, pygmy goats can be an excellent choice. These miniature dairy goats stand just 16-24 inches tall at the shoulder and weigh between 65-95 pounds. While their milk production is lower than larger breeds, pygmy goats can still provide a steady supply of high-quality milk for a small family. Pygmies are known for their friendly, docile temperament and can make delightful companions in addition to dairy animals.

Nigerian Dwarf Goats

Similar to pygmy goats, Nigerian dwarfs are a small, compact breed that can thrive in backyard settings. These goats typically stand 20-24 inches tall and weigh 65-85 pounds. Nigerian dwarfs are prolific milk producers, with does capable of yielding up to 2 quarts of milk per day. Their milk is exceptionally rich in butterfat, making it ideal for cheesemaking and other dairy products.

When choosing a goat breed for your backyard dairy farm, consider factors such as your available space, climate, desired milk production, and personal preferences. It's also important to research the specific care and housing requirements for each breed to ensure your new goats can thrive in your backyard environment.

Practical Steps for Getting Started

1. Assess your available space and resources. Determine how many goats you can comfortably accommodate and care for.

2. Research the different goat breeds and their unique characteristics to find the best fit for your backyard farm.

3. Connect with local breeders or visit goat shows to find healthy, high-quality animals.
4. Prepare proper housing, fencing, and feeding areas for your new goats.
5. Develop a routine for daily care, including milking, feeding, and healthcare.
6. Learn about goat nutrition, breeding, and herd management to ensure your animals stay healthy and productive.
7. Invest in the necessary equipment, such as milking stands, storage containers, and cheese-making supplies.
8. Consider joining a local goat-keeping or dairy farming group to connect with experienced backyard farmers and learn from their expertise.

Exercises and Activities

1. Create a comparison chart to evaluate the pros and cons of different goat breeds based on your specific needs and preferences.
2. Design a scaled layout of your backyard to determine the optimal placement and size of your goat enclosure, feeding area, and other necessary infrastructure.

3. Practice basic goat handling and milking techniques using a stuffed animal or a friend's goat (with permission) to become more comfortable and confident.

4. Experiment with simple cheesemaking or other dairy product recipes to get a feel for the process and determine which goat milk products you enjoy the most.

5. Visit a local dairy farm or goat show to observe experienced backyard farmers in action and learn from their methods and insights.

Remember, starting a backyard dairy farm with goats is a rewarding but also a significant commitment. By carefully researching your options, preparing your space, and gradually building your knowledge and skills, you can create a thriving and sustainable backyard dairy operation that provides your family with fresh, high-quality milk and dairy products.

3.3 Sheep Breeds: Considerations for Small-Scale Dairy Farming

When venturing into small-scale dairy farming, one of the most critical decisions you'll make is choosing the right sheep breed for your operation. Dairy sheep are valued not only for their milk production but also for their hardy nature, adaptability, and suitability for various farming environments. In this comprehensive guide, we'll explore the different dairy sheep breeds, their characteristics, and how to select the ideal breed for your backyard dairy farm.

Understanding Dairy Sheep Breeds

Dairy sheep are a specialized subset of the sheep family, bred primarily for their milk production. While there are numerous sheep breeds worldwide, only a few are considered true dairy breeds. *Some of the most popular and well-suited for small-scale dairy farming include:*

1. East Friesian: Originally from Germany, the East Friesian is renowned for its exceptional milk yield, with an average of 800-1,000 liters per lactation. These sheep have a calm temperament and are adaptable to various climates.

2. *Lacaune:* Hailing from France, the Lacaune breed is known for its high-quality milk, with a butterfat content of around 7%. These sheep are hardy, have a robust constitution, and are well-suited for both grazing and intensive farming systems.

3. *Awassi:* Originating from the Middle East, the Awassi breed is prized for its superior milk production, with an average of 300-500 liters per lactation. These sheep are adaptable to harsh environments and can thrive in hot, dry climates.

4. *Assaf:* A cross between the Awassi and East Friesian breeds, the Assaf sheep combines the best traits of both, including high milk yield and excellent adaptability to diverse farming conditions.

5. *Sarda:* Originating from the Italian island of Sardinia, the Sarda breed is known for its hardy nature, longevity, and consistent milk production, making it a popular choice for small-scale dairy farmers.

When selecting a dairy sheep breed, consider factors such as milk yield, butterfat content, hardiness, adaptability to your local climate, and ease of management. It's also crucial to research the availability and accessibility of these breeds in your region, as some may be more readily available than others.

Practical Considerations for Dairy Sheep Farming

Once you've identified the most suitable dairy sheep breed for your small-scale farm, there are several practical considerations to keep in mind:

1. Housing and Fencing: Provide your sheep with a comfortable, well-ventilated, and predator-proof shelter. Ensure your fencing is sturdy enough to contain the sheep and protect them from potential threats.

2. Feeding and Nutrition: Develop a balanced feeding regimen that includes high-quality forage, hay, and a suitable concentrate mix to meet the nutritional needs of your dairy sheep. This will help optimize milk production and maintain the herd's overall health.

3. Breeding and Reproduction: Establish a breeding program that aligns with your production goals. Consider factors such as breeding seasons, gestation periods, and lamb rearing to ensure a steady supply of milk.

4. Milking and Milk Handling: Invest in a proper milking setup, including a milking stand and milk storage equipment. Adhere to strict sanitation protocols to maintain milk quality and safety.

5. Health and Veterinary Care: Work closely with a veterinarian to develop a comprehensive health management plan, including vaccination schedules, parasite control, and prompt treatment of any health issues.

6. Record-Keeping: Meticulously maintain records of your dairy sheep's health, breeding, milk production, and other key performance indicators. This data will help you make informed decisions and optimize your small-scale dairy operation.

Exercises and Practical Steps

To help you get started with dairy sheep farming, consider the following exercises and practical steps:

1. Research and Visits: Visit local dairy sheep farms, attend industry events, and research online resources to gain a deeper understanding of dairy sheep breeds, their characteristics, and best management practices.

2. Needs Assessment: Evaluate your available resources, such as land, facilities, and labor, to determine the optimal herd size and dairy sheep breed that can thrive in your specific farming environment.

3. *Hands-On Experience:* Consider volunteering or interning at a successful dairy sheep farm to acquire practical, hands-on experience in all aspects of dairy sheep management.

4. *Budgeting and Financial Planning:* Develop a comprehensive business plan that includes detailed financial projections, start-up costs, and ongoing operational expenses to ensure the viability of your small-scale dairy sheep farm.

5. *Networking and Mentorship:* Connect with experienced dairy sheep farmers, industry associations, and extension services to build a support network and access valuable advice and guidance.

By carefully considering the different dairy sheep breeds, their unique characteristics, and the practical aspects of small-scale dairy farming, you'll be well-equipped to establish a thriving backyard dairy operation that provides you with a steady supply of high-quality milk and dairy products.

Chapter 3

Housing and Facilities

4. Designing a Functional and Comfortable Shelter

4.1 Housing Requirements for Dairy Animals
As a backyard dairy farmer, providing your dairy animals with appropriate housing is crucial to their health, productivity, and overall well-being. A properly designed shelter not only ensures the comfort and safety of your herd but also contributes to the efficient management of your dairy operation. In this comprehensive guide, we will explore the essential housing requirements for dairy animals, offering practical steps and exercises to help you create a functional and comfortable shelter tailored to your needs.

Understanding Dairy Animal Housing Needs
Dairy animals, such as cows, goats, or sheep, have specific requirements when it comes to

their living environment. These requirements are influenced by factors such as the animal's size, breed, age, and the local climate. By understanding these needs, you can design a shelter that effectively addresses the physical and behavioral needs of your dairy herd.

Space Requirements

Dairy animals require adequate space to move freely, rest, and access feed and water. The minimum space requirements vary depending on the size of the animal and the number of animals in the herd. *As a general guideline, allow for:*

- *Cows:* 80-120 square feet of bedded area per animal
- *Goats:* 30-50 square feet of bedded area per animal
- *Sheep:* 20-40 square feet of bedded area per animal

In addition to the bedded area, you'll need to allocate space for feed and water troughs, as well as any necessary equipment or storage.

Ventilation and Air Quality

Proper ventilation is essential to maintain good air quality and prevent the buildup of harmful gases, such as ammonia and carbon dioxide. Adequate airflow helps regulate temperature, humidity, and the removal of dust and odors, all of which can have a significant impact on the animals' health and well-being.

Temperature and Humidity Control

Dairy animals are sensitive to extreme temperatures and can experience stress if the shelter's temperature and humidity levels are not within their comfort zone. The ideal temperature range for dairy animals is typically between 40°F and 75°F, with a relative humidity between 50% and 70%.

Bedding and Flooring

The type of bedding and flooring you choose can greatly impact the comfort and cleanliness of your dairy animals' living environment. Absorbent bedding materials, such as straw, sawdust, or sand, can help maintain a dry and comfortable resting area. The flooring should be slip-resistant and easy to clean, with proper drainage to prevent the buildup of moisture and waste.

Designing a Functional and Comfortable Shelter

Now that you understand the essential housing requirements for dairy animals, let's explore the steps to designing a practical and comfortable shelter.

Step 1: Assess Your Needs

Start by evaluating the size of your herd, the available space, and the local climate. Consider factors such as the number of animals, their size, and any future expansion plans. This will help you determine the appropriate size and layout of your shelter.

Step 2: Choose the Shelter Type

Dairy animal shelters can take various forms, such as barns, sheds, or even repurposed structures. Consider the pros and cons of each option, taking into account factors like cost, ease of construction, and the specific needs of your herd.

Step 3: Plan the Shelter Layout

Divide the shelter into distinct zones for sleeping, feeding, and movement. Ensure that the layout allows for efficient flow and easy access to resources, such as feed, water, and milking equipment.

Step 4: Incorporate Ventilation and Climate Control

Design the shelter with proper ventilation systems, such as windows, vents, or fans, to maintain good air quality and regulate temperature and humidity. Consider the use of insulation, heating, or cooling systems, depending on your local climate.

Step 5: Select Appropriate Flooring and Bedding

Choose a durable, slip-resistant flooring material that is easy to clean and maintain. Provide ample bedding to ensure a comfortable resting area for your dairy animals.

Step 6: Include Necessary Amenities
Incorporate features that support the daily management of your dairy herd, such as feed and water troughs, milking stations, and storage areas for equipment and supplies.

Step 7: Implement Biosecurity Measures
Incorporate biosecurity measures to protect your herd from disease, such as dedicated entrances, cleaning and disinfection protocols, and separation of sick or new animals.

Practical Exercises and Examples
To help you put these principles into practice, here are some exercises and examples:

1. Calculating Space Requirements: Determine the total bedded area required for your herd based on the number and size of your dairy animals. Use the guidelines provided earlier to calculate the necessary space.

2. *Ventilation Evaluation:* Assess the current ventilation in your shelter or proposed shelter design. Identify any areas that may require additional airflow, and explore solutions such as strategic placement of windows, vents, or fans.

3. *Bedding and Flooring Selection:* Research different bedding materials and flooring options that would be suitable for your dairy animals. Evaluate factors like absorbency, durability, and ease of maintenance.

4. *Layout Design:* Sketch out a floor plan for your dairy animal shelter, incorporating the recommended zones for sleeping, feeding, and movement. Ensure efficient flow and access to resources.

5. *Biosecurity Checklist:* Create a checklist of biosecurity measures you can implement in your dairy animal shelter, such as dedicated entrances, cleaning protocols, and isolation areas.

Remember, the design and implementation of your dairy animal shelter should be an ongoing process, as you may need to make adjustments based on the evolving needs of your herd and changes in your operation.

Providing a functional and comfortable shelter for your dairy animals is an essential aspect of successful backyard dairy farming. By understanding the housing requirements, designing an appropriate shelter, and implementing practical measures, you can create a safe and nurturing environment that supports the health, productivity, and well-being of your dairy herd. With this comprehensive guide, you now have the knowledge and tools to embark on your journey towards creating the perfect dairy animal sanctuary.

4.2 Constructing a Milking Parlor and Handling Facility

In the world of backyard dairy farming, the milking parlor and handling facility are essential components that contribute to the overall success and well-being of your herd. These structures not only provide a safe and efficient environment for the milking process but also ensure the comfort and care of your dairy animals. In this comprehensive guide, we will delve into the intricacies of constructing a milking parlor and handling facility, equipping you with the knowledge and practical steps to create a functional and comfortable shelter for your backyard dairy operation.

Understanding the Milking Parlor

The milking parlor is the designated area where you will regularly milk your dairy animals, whether they are cows, goats, or other ruminants. This space should be designed with the comfort and safety of both the animals and the farmer in mind. A well-designed milking parlor can streamline the milking process, improve milk quality, and enhance the overall efficiency of your backyard dairy operation.

Choosing the Right Layout

When it comes to the layout of your milking parlor, there are several common configurations to consider:

1. Parallel Parlor: In this setup, the animals are positioned side by side, facing the same direction. This layout allows for easy access to the animals and can accommodate a larger number of animals in a relatively compact space.

2. Herringbone Parlor: Inspired by the pattern of herringbone fabric, this design features animals positioned at angles, forming a V-shape. This layout provides better visibility and access for the farmer, making the milking process more efficient.

3. Tandem Parlor: In a tandem parlor, the animals are positioned one behind the other, facing the same direction. This setup is often used in smaller operations and can be more suitable for handling a smaller number of animals.

Evaluate your herd size, available space, and personal preferences to determine the most suitable layout for your backyard dairy farm.

Milking Equipment and Accessories
Equipping your milking parlor with the right equipment and accessories is crucial for ensuring a smooth and efficient milking process. ***Some essential components to consider include:***

1. Milking Stanchions: These are the individual stalls where the animals will stand during the milking process. They should be designed to provide comfort and safety for the animals.

2. Milking Machines: Depending on the size of your herd, you may opt for a portable or stationary milking machine to automate the milking process.

3. Milk Storage and Cooling System: Ensure that you have the necessary equipment to store and cool the milk, maintaining its quality and freshness.

4. Cleaning and Sanitation Supplies: Maintain a clean and hygienic environment by having the appropriate cleaning and sanitizing products on hand.

Constructing the Handling Facility

In addition to the milking parlor, the handling facility is an essential component of your backyard dairy operation. This area is designed to facilitate the safe and efficient handling of your dairy animals for various purposes, such as veterinary check-ups, hoof trimming, or breeding.

Layout and Design Considerations

When planning the handling facility, consider the following factors:

1. Animal Flow: Ensure that the layout allows for a smooth and intuitive flow of animal movement, minimizing stress and potential accidents.

2. Workspace and Access: Provide ample workspace for the farmer and any assistants, as well as easy access to the animals for necessary procedures.

3. Separation and Sorting: Incorporate features that allow you to separate and sort animals as needed, such as pens or holding areas.

4. Safety Features: Implement safety features, such as sturdy fencing, non-slip flooring, and overhead protection, to prioritize the well-being of both the animals and the farmers.

Handling Equipment and Accessories

Equip your handling facility with the necessary equipment and accessories to facilitate efficient and safe animal management. *Some key items to consider include:*

1. Chutes and Crates: These structures help guide and restrain the animals during procedures, ensuring their safety and the safety of the handlers.

2. Scales: Incorporate a weighing system to monitor the health and growth of your dairy animals.

3. Hoof Trimming Stand: Provide a dedicated area for regular hoof trimming, a crucial aspect of maintaining your animals' overall health and well-being.

4. Breeding Pen: If you plan to breed your dairy animals, include a designated breeding pen to facilitate the process.

Practical Steps for Construction

Constructing a milking parlor and handling facility requires careful planning and execution. *Here are some practical steps to guide you through the process:*

1. Assess your space and herd size: Determine the available space and the number of animals you will be housing to plan the layout and size of your structures.

2. Research and plan the design: Familiarize yourself with the different layout options and design features that cater to the needs of your dairy animals and your farming goals.

3. Obtain necessary permits and approvals: Consult with local authorities to ensure that your construction plans comply with any relevant building codes or zoning regulations.

4. Prepare the site: Clear the designated area, level the ground, and ensure proper drainage to create a stable and suitable foundation for your structures.

5. Construct the milking parlor and handling facility: Depending on your level of expertise and resources, you may choose to build the structures yourself or hire a contractor to handle the construction.

6. Install the necessary equipment and accessories: Carefully integrate the appropriate milking and handling equipment to create a functional and efficient system.

7. Test and refine: Once the construction is complete, test the facilities and make any necessary adjustments to optimize their performance and the comfort of your dairy animals.

Exercises and Practical Application

To solidify your understanding and apply the concepts presented in this guide, consider the following exercises:

1. Design your ideal milking parlor layout: Sketch or create a digital model of your preferred milking parlor configuration, taking into account the size of your herd and the available space.

2. Develop a checklist for essential milking parlor equipment: Create a comprehensive list of the necessary equipment and accessories for your milking parlor, including their specifications and estimated costs.

3. Plan the layout of your handling facility: Sketch or create a digital model of your handling facility, incorporating features such as animal flow, workspace, separation, and safety considerations.

4. Research and compare different milking machine options: Investigate the various types of milking machines available on the market, evaluating their features, performance, and compatibility with your backyard dairy operation.

5. Conduct a site assessment: Carefully examine the proposed location for your milking parlor and handling facility, taking note of any potential challenges or opportunities that may arise during the construction process.

By engaging in these practical exercises, you will deepen your understanding of the concepts presented in this guide and be better equipped to construct a functional and comfortable milking parlor and handling facility for your backyard dairy farm.

4.3 Waste Management and Environmental Considerations

When embarking on your backyard dairy farming journey, it's crucial to consider the environmental impact and implement sustainable waste management practices. Proper waste handling not only benefits the environment but also ensures the health and wellbeing of your dairy animals and the surrounding ecosystem.

Understanding Dairy Waste

Dairy farming generates various types of waste, *including:*

- *Manure:* This is the primary waste product from your dairy animals and can include feces, urine, and bedding material.
- *Wastewater:* This includes water used for cleaning, milk processing, and other dairy operations.
- *Expired or spoiled dairy products:* These may include milk, cheese, or other dairy goods that have reached the end of their shelf life.

Improper management of these waste streams can lead to environmental issues such as water pollution, soil contamination, and the potential spread of diseases.

Sustainable Waste Management Strategies

To design a functional and comfortable shelter while prioritizing environmental considerations, implement the following strategies:

1. Manure Management:

- Construct a well-designed manure storage system, such as a covered concrete or steel tank, to contain the waste and prevent runoff.

- Regularly remove manure from the dairy barn and pastures, and apply it to your cropland as a natural fertilizer.

- Consider installing a composting system to transform the manure into a nutrient-rich soil amendment.

2. *Wastewater Treatment:*

- Implement a wastewater treatment system, such as a septic tank or constructed wetland, to remove contaminants and ensure proper disposal or reuse of the water.

- Direct wastewater away from streams, ponds, or other water bodies to prevent pollution.

- Consider using the treated wastewater for irrigation or other non-potable purposes.

3. *Dairy Product Disposal:*

- Develop a plan for the proper disposal of expired or spoiled dairy products, ensuring they do not end up in landfills or waterways.

- Consider donating or repurposing edible but unsold dairy products to local food banks or animal feed producers.

- Properly dispose of inedible dairy waste through appropriate channels, such as rendering facilities or licensed waste management services.

4. Energy Efficiency and Renewable Energy:
- Incorporate energy-efficient design elements in your dairy shelter, such as insulation, efficient lighting, and ventilation systems.
- Explore the use of renewable energy sources, such as solar panels or wind turbines, to power your dairy operations and reduce your carbon footprint.

5. Biodiversity and Habitat Conservation:
- Ensure that your dairy shelter and surrounding land use practices do not negatively impact local wildlife and their habitats.
- Consider planting native vegetation and creating wildlife corridors to support biodiversity.
- Avoid the use of harmful pesticides or chemicals that could contaminate the soil and water.

By implementing these waste management and environmental strategies, you can design a functional and comfortable dairy shelter that minimizes the impact on the local ecosystem and promotes sustainable dairy farming practices.

Practical Steps and Exercises

1. Conduct a Waste Audit: Assess the types and quantities of waste generated by your dairy operation. This will help you determine the appropriate waste management strategies.

2. Develop a Manure Management Plan: Create a detailed plan for the collection, storage, and application of manure on your cropland. Ensure compliance with local regulations and best practices.

3. Construct a Wastewater Treatment System: Research and install a suitable wastewater treatment system, considering the size of your dairy operation and local environmental regulations.

4. Implement a Dairy Product Disposal Procedure: Establish a system for the proper disposal or repurposing of expired or spoiled dairy products, working with local waste management services or food donation programs.

5. Explore Renewable Energy Options: Conduct an energy audit and assess the feasibility of incorporating solar, wind, or other renewable energy sources into your dairy shelter design.

6. Enhance Biodiversity: Plant native vegetation, create wildlife corridors, and implement sustainable land management practices to support local ecosystems.

By following these practical steps and exercises, you can design a functional and comfortable dairy shelter that prioritizes waste management and environmental sustainability, ensuring the long-term viability and responsible stewardship of your backyard dairy farming operation.

Chapter 4

Feeding and Nutrition

5. Nutritional Needs of Dairy Animals

5.1 Understanding Basic Nutrient Requirements

As a budding backyard dairy farmer, one of the most crucial aspects you need to understand is the nutritional needs of your dairy animals. Providing your cows, goats, or other dairy livestock with a balanced and appropriate diet is essential for their overall health, productivity, and longevity. In this comprehensive guide, we'll delve into the fundamental nutrient requirements for dairy animals and equip you with the knowledge to create an optimal feeding regimen for your backyard herd.

The Basics of Dairy Animal Nutrition

Dairy animals, like any other livestock, require a diverse array of essential nutrients to thrive.

These nutrients can be broadly categorized into six main groups: *carbohydrates, proteins, fats, vitamins, minerals,* and *water.* Each of these nutrient groups plays a crucial role in supporting the various physiological functions of the animal, from growth and lactation to immune system function and reproduction.

Carbohydrates: Carbohydrates are the primary source of energy for dairy animals. They are typically derived from sources such as grains, forages, and hay. The two main types of carbohydrates are structural carbohydrates (cellulose and hemicellulose) and non-structural carbohydrates (starches and sugars). Proper balance and source of carbohydrates are crucial for rumen health and overall productivity.

Proteins: Proteins are the building blocks of the animal's body, responsible for tissue growth, repair, and maintenance. Dairy animals require a steady supply of high-quality proteins, which can be obtained from sources like soybean meal, alfalfa, and other protein-rich feedstuffs.

Fats: Fats, or lipids, provide concentrated energy and support various bodily functions, including hormone production, cell membrane structure, and the absorption of fat-soluble vitamins. Appropriate levels of dietary fats are essential for dairy animals, particularly during high-production periods.

Vitamins: Vitamins are organic compounds required in small quantities to support a wide range of metabolic processes. Dairy animals need both fat-soluble vitamins (A, D, E, and K) and water-soluble vitamins (B-complex and C) to maintain optimal health and productivity.

Minerals: Minerals are inorganic elements that play crucial roles in bone development, enzyme function, immune system support, and more. The major minerals required by dairy animals include calcium, phosphorus, magnesium, sodium, and potassium, while trace minerals like copper, zinc, and selenium are also essential.

Water: Water is the most essential nutrient for dairy animals, as it is involved in virtually every bodily function. Ensuring that your dairy herd has access to clean, fresh water at all times is crucial for their well-being and production.

Determining Nutrient Requirements

The specific nutrient requirements for dairy animals can vary depending on factors such as age, stage of lactation, body condition, and production level. To ensure that your herd's needs are met, it's important to consult reliable resources, such as the Nutrient Requirements of Dairy Cattle published by the National Research Council (NRC), or work closely with a local veterinarian or animal nutritionist.

Here are some general guidelines for the nutrient requirements of dairy animals:

Calves and Heifers:
- *Protein:* 13-16% of the total diet
- *Energy:* 59-65 Mcal of net energy for growth per day

- ***Calcium and Phosphorus:*** 0.65-0.90% and 0.45-0.60% of the diet, respectively

Lactating Cows:
- ***Protein:*** 15-18% of the total diet
- ***Energy:*** 1.54-1.76 Mcal of net energy for lactation per pound of dry matter intake
- ***Calcium and Phosphorus:*** 0.73-0.78% and 0.45-0.60% of the diet, respectively

Goats and Sheep:
- ***Protein:*** 12-18% of the total diet
- ***Energy:*** 1.2-1.6 Mcal of net energy for lactation per pound of dry matter intake
- ***Calcium and Phosphorus:*** 0.65-0.80% and 0.45-0.55% of the diet, respectively

It's important to note that these are general guidelines, and the specific nutrient requirements for your dairy animals may vary based on their individual characteristics and production goals. Consulting with a professional is highly recommended to ensure that you are providing a balanced and optimal diet for your backyard herd.

Practical Steps for Implementing Proper Nutrition

Now that you understand the basic nutrient requirements for dairy animals, let's explore some practical steps you can take to ensure your herd is receiving the appropriate nutrition:

1. Forage-based Diet: Forages, such as pasture, hay, and silage, should make up the foundation of your dairy animals' diet. These roughage sources provide the necessary fiber, carbohydrates, and other essential nutrients.

Exercise: Conduct a forage analysis to determine the nutrient composition of your pasture or hay. Use this information to plan your supplementation strategy.

2. Balanced Concentrate Supplementation: Concentrate feeds, such as grains, protein supplements, and mineral mixes, should be used to complement the forage-based diet and provide additional energy, protein, and specific nutrient requirements.

Exercise: Calculate the amount of concentrate feed needed to balance the nutrient profile of your dairy animals' diet based on their individual requirements and the forage analysis.

3. Proper Mineral and Vitamin Supplementation: Ensure that your dairy animals are receiving the necessary vitamins and minerals through a well-formulated mineral supplement or a complete mixed ration.

Exercise: Develop a mineral and vitamin supplementation plan for your herd, taking into account factors such as stage of lactation, production levels, and environmental conditions.

4. Clean, Fresh Water: Provide your dairy animals with a constant supply of clean, fresh water. This is essential for maintaining hydration, digestion, and overall health.

Exercise: Regularly monitor and maintain your water sources to ensure they are free from contaminants and easily accessible for your dairy herd.

5. *Feeding Management:* Implement a consistent feeding schedule and ensure that feed is always available in sufficient quantities. Proper feed storage and handling can also help preserve the quality and nutritional value of your feedstuffs.

Exercise: Create a detailed feeding plan that outlines the timing, amounts, and types of feed for your dairy animals based on their individual needs and production goals.

By following these practical steps and continuously monitoring your dairy animals' performance, you can ensure that their nutritional needs are being met and optimize their health, productivity, and overall well-being in your backyard dairy farming endeavor.

5.2 Forage Management and Pasture Grazing

Proper forage management and pasture grazing are essential for ensuring the optimal health and productivity of your dairy herd. In this guide, we'll dive deep into the nutritional needs of dairy animals and explore practical strategies for managing your pastures and forages to meet those requirements.

Understanding Dairy Cow Nutritional Needs

Dairy cows have unique nutritional needs that must be carefully addressed to maintain their overall well-being and milk production. Here are the key nutrients that dairy animals require:

1. Energy: Dairy cows need a steady supply of energy to sustain their basic bodily functions, support growth and lactation, and maintain an active lifestyle. The primary sources of energy in a dairy cow's diet are carbohydrates and fats.

2. Protein: Protein is essential for the development and repair of tissues, as well as the production of milk. Dairy cows require high-quality, readily available protein sources in their diet.

3. Fiber: Fiber is crucial for proper rumen function and overall digestive health. It helps maintain a healthy pH balance in the rumen and aids in the production of butterfat in milk.

4. Minerals: Dairy cows require a balanced supply of minerals, such as calcium, phosphorus, magnesium, and trace minerals, to support bone development, immune function, and metabolic processes.

5. Vitamins: Vitamins, particularly A, D, E, and the B-complex vitamins, play vital roles in various physiological processes, including reproduction, immune function, and milk synthesis.

6. Water: Water is essential for all bodily functions and must be available in ample quantities, especially for high-producing dairy cows.

Pasture Management for Optimal Nutrition
Properly managed pastures can be an excellent source of high-quality forage to meet the nutritional needs of your dairy herd. *Here are some key considerations for effective pasture management:*

1. Grass and Legume Selection: Choose a diverse mix of grasses and legumes that thrive in your local climate and soil conditions. This will ensure a balanced and nutrient-rich forage supply throughout the grazing season.

2. Rotational Grazing: Implement a rotational grazing system, where you divide your pasture into several smaller paddocks and move your herd to a new paddock regularly. This allows the grazed areas to rest and regrow, promoting better forage quality and productivity.

3. Pasture Maintenance: Regularly mow, harrow, and reseed your pastures to maintain a healthy, dense sward and prevent the encroachment of weeds and undesirable plants.

4. Soil Fertility: Monitor your soil's nutrient levels and apply appropriate fertilizers, lime, or other soil amendments to maintain optimal soil health and fertility for robust forage growth.

5. Grazing Management: Carefully manage the grazing pressure and duration to ensure that your cows are able to consume the forage efficiently without overgrazing or damaging the sward.

Supplemental Feeding and Forage Conservation

While well-managed pastures can provide a significant portion of your dairy herd's nutritional requirements, you may need to supplement their diet with additional forages, concentrates, or other feedstuffs to ensure they meet their full nutritional needs. **Some key considerations include:**

1. Hay and Silage Production: Harvest and store high-quality hay and silage to use as supplemental feed during the non-grazing season or periods of pasture scarcity.

2. Concentrates and Supplements: Provide a balanced concentrate ration and appropriate mineral and vitamin supplements to complement the forage-based diet and address any nutritional deficiencies.

3. Feed Storage and Handling: Properly store and handle all feedstuffs to maintain their nutritional quality and prevent spoilage or contamination.

Practical Exercises and Examples

1. Forage Analysis: Collect and submit samples of your pasture forage and stored feeds for a comprehensive nutrient analysis. Use the results to formulate a balanced diet for your dairy herd.

2. Pasture Mapping and Rotation: Divide your pasture into several paddocks and create a rotational grazing plan. Monitor the growth and recovery of the forage in each paddock to optimize your grazing schedule.

3. Feeding Calculations: Determine the daily dry matter intake and nutrient requirements for your dairy cows based on their age, weight, and stage of lactation. Adjust your feeding program accordingly.

4. Hay and Silage Production: Develop a schedule for cutting, drying, and storing hay. Learn the proper techniques for making high-quality silage to use as a supplemental feed.

5. Mineral and Vitamin Supplementation: Consult with a veterinarian or livestock nutritionist to formulate a custom mineral and vitamin supplement tailored to the specific needs of your dairy herd.

By implementing these strategies and practices, you can effectively manage your pastures and forages to meet the unique nutritional needs of your dairy animals, ensuring their optimal health, productivity, and well-being.

5.3 Supplementing Feed and Ensuring a Balanced Diet

As a backyard dairy farmer, ensuring your dairy animals receive a balanced and nutritious diet is crucial for their health, productivity, and overall wellbeing. Dairy animals have specific nutritional requirements that must be met to maintain optimal milk production, growth, and reproductive performance. In this comprehensive guide, we will explore the importance of supplementing feed and creating a balanced diet for your dairy herd.

Understanding Dairy Animal Nutritional Needs

Dairy animals, such as cows, goats, and sheep, require a diverse range of nutrients to thrive. ***These include:***

1. Energy: Dairy animals need adequate energy sources, primarily from carbohydrates and fats, to support their basic metabolic processes, body maintenance, and milk production.

2. Protein: Protein is essential for growth, muscle development, and milk production. Dairy animals require high-quality protein sources, such as legumes, grains, and protein supplements.

3. Minerals: Key minerals like calcium, phosphorus, magnesium, and trace minerals (e.g., copper, zinc, selenium) are crucial for bone development, immune function, and overall health.

4. Vitamins: Vitamins, such as vitamins A, D, E, and the B-complex, play vital roles in vision, bone health, immune function, and energy metabolism.

5. *Water:* Providing clean, fresh water is essential for maintaining hydration, digestion, and overall physiological processes.

Assessing Forage and Pasture Quality

Forage and pasture are the foundation of a dairy animal's diet, providing a significant portion of their nutritional needs. Evaluating the quality and nutrient content of your forage and pasture is the first step in ensuring a balanced diet.

Steps to Assess Forage and Pasture Quality

1. Conduct a soil test to determine the nutrient composition and pH of your pasture or forage land.

2. Regularly sample and analyze the forage or pasture to determine its nutritional value, including protein, energy, and mineral content.

3. Observe the growth, color, and maturity of the forage or pasture to assess its overall quality.

4. Adjust grazing and harvesting practices to maintain optimal forage and pasture nutritional value.

Supplementing the Diet

Even with high-quality forage and pasture, dairy animals may require additional supplementation to meet their specific nutritional needs. ***Common supplemental feeds include:***

1. Concentrates: Grains, such as corn, barley, or oats, provide a concentrated source of energy and protein.

2. Protein Supplements: Soybean meal, cottonseed meal, or alfalfa hay can boost the protein content of the diet.

3. Mineral and Vitamin Supplements: Specialized mineral and vitamin mixes help ensure all essential nutrients are provided.

4. Roughages: Hay, silage, or haylage can add fiber and complement the forage-based portion of the diet.

Balancing the Diet

Balancing the diet for dairy animals involves carefully considering the nutritional requirements of the herd and adjusting the feed ration accordingly. ***This process includes:***

1. Determining the specific nutrient requirements based on factors like age, breed, lactation stage, and production goals.
2. Calculating the appropriate amounts of each feed ingredient to meet the herd's nutritional needs.
3. Monitoring the herd's body condition, milk production, and overall health to make necessary adjustments to the diet.
4. Consulting with a livestock nutritionist or extension agent to develop a customized feeding plan for your dairy herd.

Practical Steps and Exercises

1. Forage and Pasture Quality Assessment: Conduct a soil test and collect forage samples for nutrient analysis. Observe the growth and appearance of your pasture or forage land.

2. Supplemental Feed Formulation: Based on the nutrient analysis of your forage and pasture, create a supplemental feed ration that addresses any nutrient deficiencies.

3. Body Condition Scoring: Learn how to perform body condition scoring on your dairy animals to monitor their nutritional status.

4. Feeding Record Keeping: Maintain detailed records of the feed rations, quantities, and any observed changes in the herd's health or productivity.

5. Feeding Schedule and Routine: Establish a consistent feeding schedule and routine to ensure your dairy animals receive their balanced diet.

Remember, the specific nutritional needs of your dairy herd may vary based on factors such as breed, age, lactation stage, and environmental conditions. Consulting with a livestock nutritionist or experienced dairy farmer in your area can provide valuable insights and guidance to optimize the nutritional management of your backyard dairy operation.

Chapter 5

Milking and Milk Handling

6. Milking Techniques and Equipment

6.1 Proper Milking Practices for Quality Milk Production

Milking your dairy animals correctly is crucial for ensuring the production of high-quality milk, maintaining animal health, and achieving a rewarding experience in backyard dairy farming. This guide covers everything you need to know about proper milking practices, from preparation to execution and post-milking care. Let's dive into the essential steps, illustrated examples, practical exercises, and useful tips to guide you on your journey from novice to expert.

Understanding the Basics of Milking

Milking is the process of extracting milk from dairy animals, such as cows, goats, or sheep. ***Proper milking practices are essential to:***

- *Ensure milk quality and safety.*
- *Maintain the health and comfort of the animals.*
- *Prevent infections, such as mastitis.*
- *Maximize milk yield.*

Preparing for Milking

1. Create a Clean Environment:
   - *Clean the Milking Area:* Ensure the milking area is clean and dry. Remove any manure, bedding, or debris to minimize contamination.
   - *Sanitize Equipment:* Use hot water and a mild detergent to wash all milking equipment, including buckets, teat cups, and milking machines. Rinse thoroughly and sanitize with a dairy-safe sanitizer.

2. Prepare the Animal:
   - *Inspect the Animal:* Check for any signs of illness or udder abnormalities, such as swelling or redness.

- ***Wash the Udder:*** Use warm water and a clean cloth to gently wash the udder and teats. Dry thoroughly with a separate clean cloth to avoid irritation and prevent bacteria from entering the milk.

3. Prepare Yourself:
- ***Wash Your Hands:*** Use soap and warm water to thoroughly wash your hands before and after milking.
- ***Wear Clean Clothing:*** Wear clean clothes or a dedicated milking apron to reduce the risk of contamination.

The Milking Process

1. Stimulate Milk Letdown:
- ***Massage the Udder:*** Gently massage the udder and teats for a few minutes to stimulate milk letdown. This mimics the natural action of a nursing calf or kid and helps ensure a steady flow of milk.

2. Hand Milking Technique:

- *Grip the Teat:* Wrap your thumb and forefinger around the base of the teat, creating a seal.

- *Squeeze and Release:* Squeeze your fingers together while maintaining the seal, then release. Repeat this motion rhythmically to extract milk. Start with a slow, gentle squeeze to avoid discomfort to the animal.

3. Machine Milking Technique

- *Attach the Teat Cups:* If using a milking machine, attach the teat cups securely to each teat. Ensure a good seal to prevent air leaks.

- *Monitor the Machine:* Watch the milk flow and adjust the vacuum pressure as needed to maintain a steady flow. Be mindful of the animal's comfort and stop if you notice any signs of distress.

- *Detach Carefully:* Once milk flow slows down significantly, gently detach the teat cups to avoid damaging the teats.

Post-Milking Care

1. Disinfect the Teats:

- *Teat Dip:* Use a commercial teat dip solution to disinfect the teats immediately after milking. This helps prevent infections like mastitis. Dip each teat thoroughly and let the solution dry naturally.

2. Cool the Milk:

- *Immediate Cooling:* Transfer the milk to a clean container and place it in a refrigerator or an ice bath to cool it quickly. Rapid cooling preserves the quality and safety of the milk.

3. Clean the Equipment:

- *Wash and Sanitize:* Clean all milking equipment thoroughly with hot water and detergent. Rinse and sanitize to prepare for the next milking session.

Practical Steps and Exercises

To ensure you master proper milking practices, try these practical steps and exercises:

1. Practice Udder Massage:
- Spend a few minutes each day massaging the udder to get a feel for the right pressure and technique.

2. Simulate Hand Milking:
- Use a rubber glove filled with water to practice the hand milking motion. This helps develop the right grip and rhythm before milking your animal.

3. Equipment Handling:
- Familiarize yourself with milking equipment. Practice assembling, using, and disassembling milking machines to ensure smooth operation during actual milking.

Tips for Consistent Milk Quality

1. Routine and Consistency:
- Milk your animals at the same time each day to establish a routine. Consistency reduces stress for the animals and helps maintain a regular milk supply.

2. Healthy Diet:
   - Provide a balanced diet rich in nutrients to support milk production. Quality feed, fresh water, and adequate forage are essential.

3. Animal Comfort:
   - Ensure your animals have a comfortable living environment. Proper bedding, shelter, and space contribute to their overall well-being and milk quality.

4. Regular Health Checks:
   - Conduct regular health checks and maintain a vaccination schedule. Healthy animals produce better milk and are less prone to infections.

Mastering proper milking practices is a rewarding aspect of backyard dairy farming. By creating a clean environment, preparing the animal and yourself, using the correct milking techniques, and caring for the milk and equipment post-milking, you can ensure high-quality milk production. With practice and dedication, you can confidently progress from a novice to an expert in dairy farming.

Remember, each step you take towards improving your milking practices not only enhances milk quality but also promotes the health and happiness of your dairy animals. Happy milking!

6.2 Selecting and Maintaining Milking Equipment

One of the most critical aspects of backyard dairy farming is selecting and maintaining the right milking equipment. Proper equipment ensures the health of your animals, the quality of the milk, and the efficiency of your operation. This guide will help you navigate through the essentials of choosing and caring for your milking tools.

Why Proper Milking Equipment is Essential

Before diving into the specifics, let's understand why having the right milking equipment is vital:

1. Animal Health: Proper equipment minimizes the risk of injuries and infections like mastitis in your dairy animals.

2. Milk Quality: Clean, well-maintained equipment ensures that the milk remains uncontaminated and safe for consumption.

3. Efficiency: Using the right tools can save time and effort, making your dairy farming experience more enjoyable and productive.

Selecting Milking Equipment

Choosing the right milking equipment involves several steps and considerations:

1. Manual vs. Machine Milking

Manual Milking:
- ***Pros:*** Cost-effective, simple to use, no need for electricity, and good for a small number of animals.
- ***Cons:*** Time-consuming, physically demanding, and less efficient for large herds.

Machine Milking:

- **Pros:** Saves time, reduces physical effort, more efficient for larger herds, and ensures consistent milking.
- **Cons:** Higher initial cost, requires maintenance, and needs electricity or battery power.

Example Decision:
If you have one or two cows, manual milking might suffice. However, if you plan to expand, investing in a small, portable milking machine could be beneficial.

2. Types of Milking Machines

Portable Milking Machines:
- Ideal for small to medium-sized farms.
- Easy to move and clean.
- Typically includes a vacuum pump, teat cups, and a milk storage container.

Pipeline Milking Systems:
- Suitable for larger operations.

- Fixed system that transports milk directly from the cow to a bulk tank.
- More expensive and requires a dedicated milking parlor.

Example Setup:
For a backyard dairy farm with 5-10 cows, a portable milking machine is usually the best option. Brands like DeLaval, Simple Pulse, and Dansha Farms offer reliable machines for small-scale operations.

Practical Steps to Choose the Right Equipment

1. Assess Your Needs: Consider the number of animals, your budget, and future expansion plans.

2. Research Brands and Models: Look for reputable brands known for quality and durability.

3. Consult Other Farmers: Join local dairy farming groups or online forums to get recommendations and reviews.

4. Consider After-Sales Support: Ensure the brand offers good customer service and easy access to spare parts.

Maintaining Milking Equipment

Maintenance is crucial for the longevity of your equipment and the safety of your milk. *Here's a step-by-step guide to keep your milking equipment in top condition*:

1. Daily Cleaning

After Each Milking Session:
- *Rinse:* Immediately rinse all parts with lukewarm water to remove milk residues. Avoid hot water initially, as it can cook the milk proteins onto the surfaces.
- *Wash:* Use a dairy-approved detergent and hot water to thoroughly wash all components. Brushes can help scrub teats, hoses, and containers.
- *Sanitize:* After washing, rinse with a sanitizing solution to kill any remaining bacteria.

- *Dry:* Allow all parts to air dry completely before the next use to prevent bacterial growth.

2. Weekly Maintenance

- *Inspect for Wear and Tear:* Check rubber parts like teat cup liners and hoses for cracks or signs of wear. Replace them if necessary.
- *Deep Clean:* Perform a more thorough cleaning of the vacuum pump and other machine parts. Disassemble parts as needed for a deep clean.
- *Lubricate Moving Parts:* If your machine has moving parts, ensure they are well-lubricated according to the manufacturer's instructions.

3. Monthly and Annual Checks

- *Professional Servicing:* Consider having a professional service your milking machine annually to check for issues you might miss.
- *Replace Parts:* Follow the manufacturer's recommendations for replacing parts like filters,

liners, and seals regularly to maintain optimal performance.

Practical Exercises

1. Create a Cleaning Schedule: Write out a daily, weekly, and monthly cleaning and maintenance schedule. Stick it up in your milking area as a reminder.

2. Training Session: If you have helpers, conduct a training session to demonstrate the proper cleaning and maintenance techniques.

3. Equipment Audit: Once a month, conduct a thorough audit of your milking equipment to identify any issues early.

Common Problems and Solutions

1. Low Vacuum Pressure:
  - *Problem:* Milk flow is slow or inconsistent.
  - *Solution:* Check for leaks in hoses, clean or replace filters, and ensure the vacuum pump is properly maintained.

2. Milk Residue Build-Up:

- *Problem:* Milk residue is not coming off completely during cleaning.

- *Solution:* Increase the frequency of deep cleaning and ensure you're using the correct detergent and sanitizing solution.

3. Teat Injuries:

- *Problem:* Cows have irritated or injured teats after milking.

- *Solution:* Check the condition of the teat cups and liners. Ensure they are the correct size and not causing excessive suction or friction.

Selecting and maintaining the right milking equipment is a cornerstone of successful backyard dairy farming. By investing in quality tools and following a diligent maintenance routine, you can ensure the health of your animals, the quality of your milk, and the efficiency of your operation. Remember, consistency and attention to detail are key. Happy milking!

6.3 Handling, Cooling, and Storing Milk Safely

Producing your own milk from a backyard dairy farm can be immensely rewarding. However, ensuring the milk is safe for consumption requires proper handling, cooling, and storage techniques. This guide will walk you through these critical steps, providing practical advice, exercises, and examples to help you become proficient in safe milk management.

Handling Milk Safely

1. Cleanliness is Key

- *Sanitize Equipment:* Always use clean and sanitized equipment for milking. This includes milking buckets, strainers, and storage containers.
  - *Exercise:* Create a sanitizing routine. Rinse equipment with warm water immediately after use, then wash with hot soapy water, and finally sanitize with a solution of one tablespoon of

bleach per gallon of water. Rinse thoroughly with clean water before the next use.

- *Clean the Animal:* Before milking, clean the udder and teats of your cow, goat, or other dairy animal.
  - *Example:* Use a clean, damp cloth to wipe down the udder and teats, and apply a mild disinfectant or udder wash. Dry the area with a clean towel to prevent water droplets from contaminating the milk.

2. Proper Milking Technique

- *Hand Milking:* If you're milking by hand, ensure your hands are clean and dry. Use a smooth, consistent motion to express the milk.
  - *Exercise:* Practice hand milking on a dry udder first to get the technique right. Position your thumb and forefinger at the base of the teat, then close your other fingers in a gentle but firm squeezing motion.

- ***Machine Milking:*** If using a milking machine, make sure it's properly cleaned and sanitized.

- ***Example:*** Follow the manufacturer's instructions for cleaning and maintenance. Ensure the suction is not too strong to avoid injuring the animal.

3. Immediate Straining

- ***Strain Milk Quickly:*** Strain the milk through a clean cheesecloth or milk strainer to remove any debris or hair.

- ***Exercise:*** Set up a straining station. Pour milk through the strainer immediately after milking to minimize the risk of contamination.

Cooling Milk Quickly

1. Rapid Cooling

- ***Importance of Cooling:*** Bacteria multiply rapidly in warm milk, so it's crucial to cool milk as quickly as possible.

- *Example:* Aim to cool milk to below 40°F (4°C) within two hours of milking.

- *Ice Bath Method:* Place the container of milk in an ice bath.
  - *Exercise:* Prepare an ice bath in your sink or a large tub. Submerge the milk container in the ice water, stirring the milk occasionally to speed up cooling.

- *Refrigerator Method:* If an ice bath isn't feasible, place the milk container in the coldest part of your refrigerator.
  - *Example:* Use a thermometer to check your refrigerator's temperature. Ensure it's set to below 40°F (4°C).

2. Use of Cooling Tanks

- *Cooling Tanks:* For larger operations, consider investing in a bulk cooling tank.
  - *Example:* These tanks are designed to cool large quantities of milk quickly and maintain a consistent cold temperature.

Storing Milk Safely

1. Proper Containers

- *Glass or Food-Grade Plastic:* Store milk in clean, food-grade containers. Glass jars or food-grade plastic jugs are ideal.
  - *Exercise:* Regularly inspect storage containers for cracks or signs of wear. Replace any damaged containers to prevent contamination.

- *Labeling:* Always label containers with the date of milking.
  - *Example:* Use waterproof markers or labels to ensure the date is clear and legible.

2. Refrigerator Storage

- *Cold Storage:* Store milk in the main body of the refrigerator, not the door, to keep it at a consistent temperature.

- *Exercise:* Organize your refrigerator to ensure milk is stored at the back, where it's coldest. Avoid overloading the fridge to allow for proper air circulation.

3. Regular Monitoring

- *Check Temperature:* Regularly check the temperature of your refrigerator.
- *Exercise:* Place a thermometer in your refrigerator and monitor it daily to ensure it stays below 40°F (4°C).

4. Shelf Life and Usage

- *Use Within a Week:* Ideally, consume fresh milk within a week.
- *Example:* Plan your dairy usage to ensure milk is consumed while it's still fresh. Make cheese, yogurt, or other dairy products if you have surplus milk.

- *Freezing Milk:* If you have excess milk, consider freezing it.

- *Exercise:* Freeze milk in small batches using ice cube trays or freezer-safe containers. Thaw in the refrigerator and use within a few days.

Practical Tips and Exercises

1. Daily Routine: Establish a daily milking and cleaning routine to maintain consistency and cleanliness.

- *Example:* Milk your animals at the same times each day. Clean and sanitize equipment immediately after each use.

2. Record Keeping: Keep detailed records of milking times, quantities, and any issues with the animals or equipment.

- *Exercise:* Start a dairy logbook. Note the date, time, and amount of milk produced each day, along with any observations about the animal's health or behavior.

3. Milk Quality Testing: Regularly test milk for quality and safety.

- ***Example:*** Perform simple tests such as checking for off-odors, unusual colors, or sourness. Consider sending samples to a lab for bacterial testing if you suspect contamination.

4. Education and Training: Continuously educate yourself about best practices in dairy farming.

- ***Exercise:*** Attend workshops, read books, and join online forums or local farming groups to stay informed about the latest techniques and regulations in dairy farming.

By following these guidelines and incorporating the suggested exercises and examples into your routine, you can ensure that the milk produced on your backyard dairy farm is handled, cooled, and stored safely. This not only guarantees high-quality, delicious milk but also protects the health of you and your family. Happy farming!

Chapter 6

Health and Disease Management

7. Maintaining Optimal Health and Preventing Diseases

7.1 Vaccination Protocols and Parasite Control

In backyard dairy farming, maintaining the health of your livestock is paramount. Just like humans, animals need regular healthcare to prevent diseases and stay in optimal condition. Two critical components of livestock health management are vaccination protocols and parasite control. This guide will delve into these topics, providing you with comprehensive guidance to ensure your dairy animals remain healthy and productive.

Understanding Vaccination Protocols

Vaccinations are vital to protect your animals from a variety of infectious diseases. These diseases can significantly impact their health, milk production, and even lead to death. Vaccination works by stimulating the animal's immune system to recognize and fight specific pathogens if they are encountered in the future.

Key Vaccinations for Dairy Animals

1. Clostridial Diseases: Includes diseases like tetanus and blackleg. Vaccinate calves around 2-4 months and give booster shots annually.
2. Bovine Respiratory Disease Complex (BRD): Often referred to as "shipping fever." Vaccinate calves at 2-4 months and booster annually.
3. Brucellosis: Also known as Bang's disease, this is crucial for heifers. Vaccinate heifers between 4-12 months old.
4. Leptospirosis: Can cause abortions and milk production issues. Annual vaccination is recommended.
5. Bovine Viral Diarrhea (BVD): Affects the respiratory and reproductive systems. Vaccinate calves at 2-4 months and booster annually.

Practical Steps for Vaccination

1. Schedule Vaccinations: Keep a calendar to track when each animal needs its vaccinations. Group vaccinations by age and type to simplify the process.

2. Use Proper Storage: Vaccines must be stored at the correct temperature, usually refrigerated, to maintain efficacy.

3. Administer Correctly: Follow the manufacturer's instructions for administration, whether subcutaneous (under the skin) or intramuscular (into the muscle).

4. Maintain Records: Keep detailed records of all vaccinations, including date, type of vaccine, and any reactions observed.

Example: Vaccinating a Calf

1. Preparation: Ensure you have the correct vaccine and syringes. Clean the injection site with alcohol.

2. Administration: If the vaccine is subcutaneous, pinch a fold of skin and inject just under the skin. If intramuscular, find a large muscle area like the neck and inject directly into the muscle.

3. Post-Vaccination: Observe the calf for any signs of adverse reactions such as swelling, fever, or lethargy.

Parasite Control

Parasites can wreak havoc on the health of your dairy animals, affecting their weight, milk production, and overall well-being. There are two main types of parasites to be aware of: *internal parasites (worms)* and *external parasites (ticks, lice, mites).*

Identifying Parasites

- *Internal Parasites:* Symptoms include weight loss, diarrhea, poor coat condition, and anemia.
- *External Parasites:* Symptoms include itching, hair loss, skin irritations, and visible parasites on the skin.

Practical Steps for Parasite Control

1. Regular Deworming: Implement a deworming schedule. For example, deworm calves at 2-4 months and then every 6 months.

2. Pasture Management: Rotate pastures to prevent overgrazing and reduce parasite loads. Avoid grazing new animals on heavily infested pastures.

3. Sanitation: Keep living areas clean and dry. Regularly clean bedding, feed areas, and water troughs to minimize parasite environments.

4. Natural Predators: Introduce beneficial insects that prey on parasites, such as certain types of beetles that eat fly larvae.

Example: Deworming a Dairy Cow

1. Choose the Right Dewormer: There are several types available, including oral drenches, injectables, and pour-ons. Select based on the parasite load and type.

2. Administer the Dewormer: For oral drenches, restrain the cow and use a dosing gun to administer the medication directly into the mouth. Ensure the cow swallows the dose.

3. Monitor: After deworming, observe the cow for any adverse reactions and improvements in health.

Exercises for Practical Application

1. Create a Vaccination Schedule: List all your animals and create a timeline for necessary vaccinations. Include booster shots and make notes of any special instructions.

2. Parasite Check Routine: Set up a weekly routine to check for external parasites. Use a fine-toothed comb to inspect for lice and mites, and part the hair to look for ticks.

3. Pasture Rotation Plan: Map out your grazing areas and develop a rotation plan that allows each section to rest and recover. Include fencing strategies to facilitate easy movement of livestock.

Maintaining the health of your dairy animals through effective vaccination protocols and parasite control is essential for a successful backyard dairy farming operation. By understanding the diseases to vaccinate against, how to administer vaccines correctly, and implementing robust parasite control measures, you can ensure your animals thrive, resulting in better milk production and overall farm productivity. Remember, consistent monitoring and record-keeping are your best tools in preventing and managing health issues.

7.2 Identifying and Treating Common Health Issues

Backyard dairy farming can be a rewarding endeavor, providing fresh milk and a closer connection to your food source. However, it also comes with responsibilities, particularly in maintaining the health of your dairy animals. This section will guide you through identifying and treating common health issues that you may encounter.

Identifying Common Health Issues

Understanding the signs of health issues in dairy animals is crucial. Here's how you can identify some of the most common problems:

1. Mastitis

- ***Symptoms:*** Swollen, hot, or hard udder; decreased milk production; milk with clots or pus.
- ***Detection:*** Regularly check the udder and milk quality. Strip cup tests can help in early detection.

2. Bloat

- ***Symptoms:*** Swelling on the left side of the animal; discomfort or restlessness; labored breathing.
- ***Detection:*** Observe the animal after feeding. Bloat often occurs when animals eat too quickly or consume too much fresh pasture.

3. Foot Rot

- ***Symptoms:*** Lameness; foul odor; swelling and redness between the toes.

- *Detection:* Regularly inspect the feet, especially in wet conditions.

4. Parasites (Internal and External)
- *Symptoms:* Weight loss; diarrhea; rough coat; scratching or rubbing.
- *Detection:* Conduct regular fecal tests and inspect the skin for lice or mites.

5. Respiratory Infections
- *Symptoms:* Coughing; nasal discharge; fever; difficulty breathing.
- *Detection:* Pay attention to changes in breathing patterns and general demeanor.

Treating Common Health Issues

Treatment can vary based on the specific health issue. *Here are practical steps to address each of the common problems:*

1. Mastitis Treatment
- *Immediate Action:* Isolate the affected animal to prevent spreading. Milk out the infected quarter frequently.

- *Medications:* Consult a vet for appropriate antibiotics.

- *Preventive Measures:* Maintain hygiene during milking, ensure proper nutrition, and avoid injuries to the udder.

2. Bloat Treatment

- *Immediate Action:* Walk the animal to help release gas. Administer anti-bloat medications (such as mineral oil or commercial anti-bloat products).

- *Emergency:* If severe, a vet may need to relieve pressure using a trocar or a stomach tube.

- *Preventive Measures:* Feed hay before allowing access to lush pasture and provide baking soda as a preventive.

3. Foot Rot Treatment

- *Immediate Action:* Clean the affected area and remove debris. Apply antiseptic solutions.

- *Medications:* Use footbaths with copper sulfate or zinc sulfate solutions.

- *Preventive Measures:* Keep the living area dry and clean. Regularly trim hooves.

4. Parasite Treatment

- *Immediate Action:* Administer dewormers or insecticides as recommended by a vet.
- *Medications:* Use broad-spectrum dewormers and topical treatments for external parasites.
- *Preventive Measures:* Rotate pastures to break the life cycle of parasites and maintain cleanliness.

5. Respiratory Infection Treatment

- *Immediate Action:* Isolate the animal to prevent spreading. Provide a warm, dry environment.
- *Medications:* Administer antibiotics as prescribed by a vet.
- *Preventive Measures:* Ensure good ventilation in housing and avoid sudden changes in weather exposure.

Practical Steps and Exercises for Health Maintenance

1. Daily Observation

- Spend time each day observing your animals for any changes in behavior, appetite, or physical condition.

- ***Exercise:*** Keep a health log for each animal to track any abnormalities and ensure timely intervention.

2. Regular Health Checks

- Conduct thorough health inspections weekly, focusing on the udder, feet, coat, and overall demeanor.

- ***Exercise:*** Create a checklist for these inspections to ensure you don't miss any critical areas.

3. Proper Nutrition

- Provide a balanced diet tailored to the nutritional needs of dairy animals, including minerals and vitamins.

- ***Exercise:*** Work with a nutritionist to develop a diet plan and adjust based on seasonal changes and animal conditions.

4. Hygiene Practices

- Maintain a clean living environment to prevent infections and parasite infestations.
- *Exercise*: Develop a cleaning schedule for barns and milking equipment. Use natural disinfectants where possible.

5. Training and Handling

- Gentle and consistent handling reduces stress, which can prevent many health issues.
- *Exercise*: Spend time training your animals to be comfortable with human contact and routine procedures.

Case Study: Treating Mastitis in a Backyard Cow

Background: You notice that one of your cows, Daisy, has a swollen and painful udder. Her milk production has also decreased, and the milk appears lumpy.

Steps Taken:

1. Isolation: You isolate Daisy from the rest of the herd to prevent the potential spread of infection.

2. Observation: A thorough inspection confirms signs of mastitis.

3. Milking Out: You milk out the infected quarter frequently to remove the bacteria.

4. Veterinary Consultation: You contact your vet, who prescribes an antibiotic treatment.

5. Hygiene: Enhanced hygiene practices during milking are implemented.

6. Follow-Up: You continue to monitor Daisy, ensuring she completes her course of antibiotics and returns to normal health.

Outcome: Daisy recovers fully, and you gain valuable experience in early detection and treatment of mastitis.

By understanding the signs of common health issues and knowing how to address them, you can ensure the well-being of your dairy animals. Regular observation, proper nutrition, and maintaining a clean environment are key to preventing many health problems. Remember, timely intervention can make a significant difference in the health and productivity of your backyard dairy farm.

By following these guidelines and staying vigilant, you'll be well on your way to becoming an expert in backyard dairy farming, ensuring your animals stay healthy and your milk supply remains safe and abundant.

7.3 Biosecurity Measures and Disease Prevention Strategies

Raising dairy animals in your backyard can be an enriching and rewarding experience. However, it comes with responsibilities, including ensuring the health and safety of your animals. Biosecurity and disease prevention are critical aspects of dairy farming, helping to protect your animals from illness, ensuring the safety of milk production, and maintaining the overall well-being of your farm. This chapter will walk you through essential biosecurity measures and disease prevention strategies, offering practical steps, examples, and exercises to solidify your understanding.

Understanding Biosecurity in Dairy Farming
Biosecurity refers to the procedures and measures taken to protect your farm from the introduction and spread of diseases. Effective biosecurity minimizes the risk of disease outbreaks, which can devastate your animals and reduce milk production.

Key Components of Biosecurity

1. Isolation and Quarantine:
   - *New Animals:* Always quarantine new animals for at least 30 days before introducing them to your herd. This period allows you to monitor for any signs of illness.
   - *Sick Animals:* Separate any animal showing signs of illness to prevent the spread of disease.

Example: You purchase a new goat. Set up a separate pen away from your existing animals and monitor it daily for signs of illness such as coughing, diarrhea, or lethargy.

2. Controlled Access:
 - *Limit Visitors:* Restrict access to your farm to only essential personnel and ensure they follow biosecurity protocols.
 - *Sanitize Footwear and Equipment:* Use disinfectant footbaths and clean equipment regularly to prevent the spread of pathogens.

Example: Place a footbath with disinfectant at the entrance of your dairy area. Require everyone to step through it before entering.

3. Personal Hygiene:
 - *Hand Washing:* Wash your hands before and after handling animals, especially before milking.
 - *Clean Clothing:* Wear clean, farm-specific clothing and change before leaving the farm to avoid carrying contaminants off-site.

Example: Keep a set of clothes and boots specifically for farm work and change into them only when you're handling your animals.

4. Pest Control:

- *Rodents and Insects:* Implement measures to control pests that can carry diseases. Use traps, barriers, and regular cleaning to minimize their presence.

Example: Set up rodent traps and regularly check them. Keep feed in sealed containers to avoid attracting pests.

Disease Prevention Strategies

Preventing disease is a combination of good management practices, regular health monitoring, and vaccination. Below are some essential strategies to keep your dairy animals healthy.

Vaccination

Vaccination is a proactive approach to disease prevention. Consult with a veterinarian to develop a vaccination schedule tailored to your specific animals and region.

Example: If you're raising dairy goats, vaccines against clostridial diseases (such as tetanus and enterotoxemia) are commonly recommended.

Regular Health Checks

Performing regular health checks helps in early detection of diseases, allowing for timely intervention.

Steps for Regular Health Checks
1. Observe Behavior: Daily observation can help spot any changes in behavior, appetite, or activity levels.
2. Physical Examination: Regularly check for signs such as abnormal discharge, swelling, lameness, or changes in body condition.

3. Record Keeping: Maintain detailed records of each animal's health status, treatments, and vaccination dates.

Example: Set aside time every morning to observe your herd as they come for feeding. Look for signs of lethargy, limping, or any animals that isolate themselves from the group.

Clean Environment

A clean environment reduces the risk of disease by minimizing the presence of harmful pathogens.

Practical Steps for Maintaining Cleanliness
1. Clean Bedding: Replace bedding regularly to keep it dry and clean.
2. Sanitize Equipment: Clean and sanitize milking equipment after each use.
3. Manure Management: Remove manure daily and compost it away from the animal housing area.

Exercise: Create a weekly cleaning schedule. For example, designate Mondays and Thursdays for deep cleaning of the animal pens, ensuring all bedding is replaced and surfaces are sanitized.

Nutrition and Water

Proper nutrition and clean water are fundamental to the health of your dairy animals.

Tips for Nutrition and Water

1. Balanced Diet: Provide a balanced diet rich in essential nutrients. Consult a nutritionist if needed.

2. Clean Water: Ensure constant access to clean, fresh water. Clean water troughs regularly to prevent contamination.

Example: Incorporate a mix of high-quality hay, grains, and a mineral supplement into your animals' diet. Check water troughs daily and scrub them out weekly.

Stress Reduction

Reducing stress in your animals helps maintain a robust immune system, making them less susceptible to diseases.

Strategies for Stress Reduction

1. Proper Housing: Provide adequate space and shelter to prevent overcrowding and exposure to harsh weather.

2. Gentle Handling: Handle animals calmly and avoid rough or sudden movements that can cause stress.

Example: Design your housing with plenty of space per animal, ensuring each has access to shelter, feed, and water without competition.

Monitoring and Record Keeping

Keeping detailed records allows you to track the health history of each animal, making it easier to spot patterns and manage outbreaks.

Key Records to Maintain

1. Health Records: Document any illnesses, treatments, and vaccinations.

2. Production Records: Keep track of milk production levels to spot any decreases that might indicate health issues.

3. Breeding Records: Record breeding dates and outcomes to monitor reproductive health.

Exercise: Set up a simple record-keeping system, either digitally or in a notebook. Log daily observations, health checks, and any treatments administered.

Practical Exercises

1. Create a Biosecurity Plan: Draft a biosecurity plan for your farm. Include protocols for quarantine, visitor access, hygiene, pest control, and regular health checks.

2. Perform a Mock Health Check: Practice a health check on one of your animals. Write down your observations and compare them with your previous records.

3. Design a Cleaning Schedule: Develop a detailed cleaning schedule for your farm, outlining daily, weekly, and monthly tasks.

4. Evaluate Your Farm Layout: Assess your farm's layout for potential biosecurity risks. Identify areas for improvement, such as adding footbaths or reorganizing quarantine areas.

By implementing these biosecurity measures and disease prevention strategies, you can create a healthy, productive, and sustainable environment for your backyard dairy farm. Taking proactive steps to ensure the well-being of your animals not only enhances their quality of life but also secures your investment and the quality of the dairy products you produce.

Chapter 7

Dairy Product Processing

8. Exploring Value-Added Dairy Products

8.1 Butter, Cheese, Yogurt, and Other Products

One of the most rewarding aspects of backyard dairy farming is the ability to transform fresh milk into a variety of delicious and nutritious dairy products. This chapter will guide you through the processes of making butter, cheese, yogurt, and other value-added products, providing practical steps, exercises, and tips to ensure success, even for beginners.

Butter

Butter is a staple in many households, and making it at home can be a simple and gratifying experience. *Here's how to do it:*

Ingredients:

- Fresh cream from your cow or goat's milk

Equipment:
- A churn (hand-cranked or electric)
- Strainer
- Salt (optional)
- Cold water

Steps:

1. Collect the Cream: Let your fresh milk sit for about 24 hours in a cool place. The cream will rise to the top. Skim this cream off gently with a ladle.

2. Churning: Pour the cream into your churn. If you don't have a churn, a stand mixer with a whisk attachment works too. Churn the cream until it thickens and separates into butter and buttermilk. This process can take 10-15 minutes.

3. Washing the Butter: Once you see the butter forming, strain out the buttermilk (save this for baking or drinking). Rinse the butter under cold water, pressing it with a wooden spoon or butter paddle to remove as much buttermilk as possible. Repeat this until the water runs clear.

4. Salting and Storing: If you prefer salted butter, now is the time to add salt and mix it in thoroughly. Shape the butter as desired and store it in the refrigerator.

Exercise:
Try making a small batch of butter using 1 cup of cream. Note how long it takes to churn and how much butter you yield. This will help you understand the process better before scaling up.

Cheese

Cheese-making can range from simple fresh cheeses to complex aged varieties. Here, we'll start with a basic cheese that's easy for beginners: ricotta.

Ingredients:
- 1 gallon of fresh milk
- 1/4 cup white vinegar or lemon juice
- Salt to taste

Equipment:
- Large pot
- Thermometer
- Cheesecloth
- Strainer
- Large bowl

Steps

1. Heating the Milk: Pour the milk into a large pot and heat it slowly to 185°F (85°C), stirring occasionally to prevent scorching.

2. Adding Acid: Once the milk reaches the desired temperature, add the vinegar or lemon juice. Stir gently, and you'll notice the curds beginning to form. Let it sit for 10 minutes.

3. Straining: Line a strainer with cheesecloth and place it over a large bowl. Pour the curdled milk into the strainer, allowing the whey to drain away. Let it drain for about 30 minutes or until it reaches the desired consistency.

4. Finishing: Transfer the ricotta to a bowl, add salt to taste, and mix well. Your fresh ricotta is now ready to eat or use in recipes.

Exercise:

Make a batch of ricotta and try it in different dishes, such as lasagna, stuffed shells, or simply spread on toast with honey.

Yogurt

Homemade yogurt is not only delicious but also packed with probiotics. The process is straightforward and can be done with minimal equipment.

Ingredients:

- 1 quart of fresh milk
- 2 tablespoons of plain yogurt with live cultures (as a starter)

Equipment:

- Pot
- Thermometer
- Whisk

- Jars or containers
- Insulated cooler or yogurt maker

Steps

1. Heating the Milk: Pour the milk into a pot and heat it to 180°F (82°C), then let it cool to 110°F (43°C).

2. Adding the Starter: Once the milk has cooled, whisk in the plain yogurt. This introduces the live cultures necessary for fermentation.

3. Incubating: Pour the mixture into jars or containers. Keep the containers warm at around 110°F (43°C) for 6-8 hours. This can be done using a yogurt maker, an insulated cooler filled with warm water, or an oven with the light on.

4. Chilling and Storing: After incubation, transfer the yogurt to the refrigerator to cool and set. You can store it for up to two weeks.

Exercise:

Experiment with different incubation times to find your preferred yogurt texture and tanginess. Try making flavored yogurt by adding fresh fruit or honey after the yogurt has set.

Other Value-Added Dairy Products

Once you're comfortable with butter, cheese, and yogurt, there are many other dairy products you can explore:

1. Cream Cheese: Similar to yogurt, but strained longer and sometimes with added flavors.

2. Kefir: A fermented milk drink that's even easier to make than yogurt. Simply add kefir grains to milk, let it ferment at room temperature for 24 hours, then strain the grains out.

3. Ghee: Clarified butter made by heating butter until the milk solids separate and can be removed. Ghee has a higher smoke point and longer shelf life than regular butter.

Practical Tips for Success

- *Quality Milk:* Always start with the freshest, highest-quality milk you can get. Raw milk is ideal for many dairy products because it hasn't been pasteurized, preserving the natural bacteria and enzymes.
- *Cleanliness:* Maintain a high standard of cleanliness to prevent contamination. Sterilize your equipment and work in a clean environment.
- *Temperature Control:* Use a thermometer to monitor temperatures accurately. Many dairy processes are sensitive to temperature changes.
- *Patience:* Some products, like aged cheese, require time and patience. Don't rush the process.

Making your own dairy products at home can be a fun and rewarding endeavor. It not only provides you with fresh, delicious food but also gives you a deeper connection to the source of your dairy. By following these steps and practicing regularly, you'll soon become proficient in transforming fresh milk into a variety of value-added products. Enjoy the journey and savor the fruits of your labor!

8.2 Safe Processing Techniques and Equipment

Backyard dairy farming is an increasingly popular pursuit, offering the joy of producing fresh milk and the satisfaction of crafting various dairy products. However, to ensure the safety and quality of these products, understanding safe processing techniques and the proper use of equipment is crucial. This guide will help you navigate the world of value-added dairy products, from novice to expert, focusing on safety and best practices.

Why Safe Processing Matters

Safe processing of dairy products is vital to prevent foodborne illnesses and ensure high-quality, delicious products. Milk is a highly perishable item, and improper handling can lead to contamination with harmful bacteria like E. coli, Salmonella, and Listeria. Following safe processing techniques minimizes these risks and preserves the nutritional value and taste of your dairy products.

Basic Equipment for Dairy Processing

Before diving into specific techniques, let's look at the essential equipment you'll need:

1. Milking Equipment: Includes stainless steel buckets, filters, and teat dip cups. Stainless steel is preferred because it's easy to clean and doesn't harbor bacteria.

2. Thermometer: For checking milk temperatures during pasteurization.

3. Pasteurizer: A device for heating milk to a specific temperature to kill harmful bacteria.

4. Cheesecloth or Butter Muslin: For straining cheese and butter.

5. Cheese Press: If you plan to make hard cheeses.

6. Butter Churn: Optional, but helpful for making butter.

7. Refrigeration: To store milk and dairy products at safe temperatures.

8. Sanitizing Solutions: For cleaning equipment thoroughly before and after use.

Practical Steps for Safe Dairy Processing

1. Milking

- ***Clean Environment:*** Ensure the milking area and the animal are clean. Wash your hands thoroughly before milking.
- ***Sterilize Equipment:*** Clean and sterilize all milking equipment before use. A simple method is using hot water and a mild detergent, followed by a sanitizing solution.
- ***Proper Milking Techniques:*** Use gentle, consistent pressure. Discard the first few squirts of milk, as they can contain bacteria.

2. Pasteurization

Pasteurization involves heating milk to a specific temperature to kill harmful bacteria. There are two common methods:

- ***Low-Temperature, Long-Time (LTLT):*** Heat milk to 145°F (63°C) and hold it for 30 minutes.
- ***High-Temperature, Short-Time (HTST):*** Heat milk to 161°F (72°C) and hold it for 15 seconds.

Steps for Pasteurization:
1. Pour milk into the pasteurizer.
2. Heat to the required temperature, stirring occasionally.
3. Maintain the temperature for the required duration.
4. Cool the milk rapidly by placing the container in an ice bath or refrigerator.

3. Making Yogurt

Yogurt is a popular value-added product that's simple to make at home.

Ingredients:
- Fresh milk
- Yogurt starter culture

Steps:

1. Heat milk to 180°F (82°C) to denature proteins, ensuring a thicker yogurt.
2. Cool the milk to 110°F (43°C).
3. Add the starter culture (a few tablespoons of pre-made yogurt or a commercial yogurt starter) to the cooled milk.
4. Pour the mixture into a clean container and incubate at 110°F (43°C) for 6-12 hours. A yogurt maker can maintain this temperature consistently.
5. Once the yogurt is set, refrigerate it to stop the fermentation process.

4. Cheese Making

Cheese making can range from simple to complex, but basic cheese like ricotta or paneer is a good start.

Ingredients:
- Fresh milk
- Acid (lemon juice or vinegar) or rennet

Steps for Ricotta:

1. Heat milk to 195°F (90°C).
2. Add acid (lemon juice or vinegar) to the milk and stir gently until curds form.
3. Let it sit for a few minutes to firm up the curds.
4. Pour the mixture through a cheesecloth-lined colander to drain the whey.
5. Gather the cheesecloth corners and hang it to drain for an additional hour.
6. Transfer to a container and refrigerate.

Exercises to Enhance Skills

1. Sanitation Practice: Set up a routine for cleaning and sanitizing your dairy equipment. Create a checklist and follow it meticulously.

2. Temperature Control: Practice using your thermometer to monitor temperatures accurately during pasteurization and yogurt making.

3. Recipe Experimentation: Try different recipes and techniques for making yogurt and cheese to understand how variations in process affect the final product.

4. Record Keeping: Keep detailed records of your processes, temperatures, and times. Note any deviations and their outcomes to improve your methods over time.

Tips for Success

- *Quality Milk:* Start with high-quality, fresh milk. The better your milk, the better your end product.
- *Consistent Hygiene:* Always prioritize cleanliness to prevent contamination.
- *Patience:* Many dairy products require time and careful attention. Don't rush the process.
- *Learning and Adaptation:* Be prepared to learn from mistakes and adjust your techniques as needed.

Safe processing techniques and the right equipment are the foundation of successful backyard dairy farming. By following the steps outlined in this guide, you'll not only ensure the safety and quality of your dairy products but also enjoy the rewarding experience of creating delicious, value-added products at home. Whether you're making yogurt, cheese, or butter, the journey from novice to expert begins with a commitment to safety, cleanliness, and continuous learning. Happy dairying!

8.3 Marketing and Selling Your Dairy Products

Backyard dairy farming can be a rewarding venture, providing fresh dairy products for your family and potentially generating income. One way to increase profitability is by creating and selling value-added dairy products. These products not only cater to diverse consumer preferences but also offer higher profit margins compared to raw milk. This guide will take you through the process of marketing and selling your dairy products, emphasizing value-added options.

Understanding Value-Added Dairy Products

Value-added dairy products are those that have undergone some form of processing, packaging, or enhancement to increase their market value. Examples include:

1. Cheese
2. Yogurt
3. Butter
4. Cream and Ice Cream
5. Ghee
6. Flavored Milk
7. Dairy-based Beverages

These products often fetch higher prices due to their convenience, unique flavors, and longer shelf life.

Benefits of Value-Added Products

1. Increased Profit Margins: Processed products often sell for more than raw milk.

2. Market Diversification: Offering a variety of products attracts different customer segments.

3. Brand Loyalty: Unique products can create a loyal customer base.

4. Reduced Waste: Processing milk into various products can minimize waste.

Steps to Start Marketing and Selling Value-Added Dairy Products

1. Research and Planning
Before diving into production, conduct thorough market research to identify demand and competition. *Ask yourself:*
- What products are popular in my area?
- What are my competitors offering?
- What are the local regulations for dairy processing and sales?

2. Develop a Business Plan
A solid business plan outlines your goals, target market, marketing strategy, and financial projections. It helps in understanding the feasibility and profitability of your venture.

3. Learn the Craft
Master the production of your chosen value-added products. *Here are some practical steps for a few products:*

- ***Cheese Making:*** Start with basic cheeses like mozzarella or cheddar. Invest in a cheese press and follow step-by-step recipes.
- ***Yogurt Production:*** Use a reliable yogurt starter culture and maintain the correct incubation temperature. Homemade yogurt can be flavored with fruits or honey.
- ***Butter Churning:*** Collect cream from your milk and use a butter churner. Add salt or herbs for variety.

4. Equip Your Dairy

Invest in necessary equipment and supplies, ***such as:***
- Pasteurizers
- Cheese presses
- Butter churners
- Yogurt makers
- Packaging materials

Ensure your dairy complies with local health and safety regulations.

5. Packaging and Branding

Attractive packaging and strong branding can differentiate your products from competitors.
Consider:
- Eye-catching labels
- Reusable containers
- Clear ingredient lists

6. Set Your Pricing

Determine your pricing based on production costs, market rates, and desired profit margins. Value-added products typically allow for higher pricing, but ensure your prices are competitive.

7. Marketing Strategies

Use various strategies to market your products
- *Local Farmers Markets:* Set up a stall to sell directly to consumers.
- *Online Sales:* Create a website or use social media platforms to reach a broader audience.

- ***Subscription Services:*** Offer weekly or monthly delivery services for fresh dairy products.
- ***Partnerships:*** Collaborate with local grocery stores or restaurants to stock your products.

Practical Exercises

1. Market Survey Exercise: Visit local farmers markets and grocery stores. Note down the types of dairy products available, their prices, and packaging styles. This will help you understand the competition and demand.

2. Product Testing Exercise: Create small batches of different value-added products. Distribute samples to friends and family, gathering feedback on taste, packaging, and pricing.

3. Branding Exercise: Design labels for your products. Consider hiring a graphic designer or using online tools like Canva to create professional-looking labels.

4. Financial Planning Exercise: Calculate the cost of production for each product, including raw materials, labor, and packaging. Determine your selling price to ensure a profitable margin.

Example: Launching a Homemade Yogurt Line

Let's walk through a practical example of launching a homemade yogurt line.

1. Market Research: You find that Greek yogurt is popular, but there are few local producers.

2. Business Plan: Outline your goals, target market (health-conscious individuals), marketing strategies (farmers markets and online), and financial projections.

3. Learning the Craft: Perfect your Greek yogurt recipe, ensuring it's thick and creamy. Experiment with different flavors like honey, blueberry, and vanilla.

4. Equipping Your Dairy: Purchase a yogurt maker and sterilized jars for packaging.

5. Branding and Packaging: Design a logo and labels that highlight the health benefits and natural ingredients of your yogurt. Use eco-friendly jars to appeal to environmentally conscious consumers.

6. Pricing: Calculate the cost per jar, considering milk, cultures, labor, and packaging. Price your yogurt competitively, perhaps slightly higher due to its homemade quality.

7. Marketing: Set up a stall at local farmers markets. Offer samples to attract customers. Create social media profiles to showcase your yogurt-making process and engage with potential customers. Consider offering a subscription service for weekly deliveries.

Creating and selling value-added dairy products can significantly enhance the profitability of your backyard dairy farming venture. By understanding market demand, mastering production techniques, and implementing effective marketing strategies, you can build a loyal customer base and achieve financial success. Remember, the key is to start small, continuously improve your products, and listen to customer feedback. Happy dairy farming!

Chapter 8

Troubleshooting and Problem-Solving

9. Overcoming Challenges in Backyard Dairy Farming

9.1 Managing Seasonal Variations and Weather Conditions

Backyard dairy farming is a rewarding endeavor, but it comes with challenges, particularly when dealing with seasonal variations and weather conditions. Understanding how to manage these fluctuations is essential to maintain the health of your livestock and the productivity of your dairy farm. This guide will help you navigate these challenges with practical steps, exercises, and examples.

Understanding Seasonal Variations

Seasonal changes can affect dairy farming in numerous ways. The primary concerns revolve around temperature extremes, changes in feed availability, and alterations in livestock behavior and health.

Summer

Challenges:
- Heat stress
- Dehydration
- Reduced milk production

Solutions:

1. Provide Shade and Ventilation:
- Set up shade structures like tarps or plant trees in the pasture.
- Ensure barns and shelters are well-ventilated. Install fans if necessary.

2. Hydration:
- Always have clean, cool water available.
- Consider adding electrolytes to the water during extremely hot days.

3. Monitor Health:

- Watch for signs of heat stress: excessive panting, reduced appetite, and lethargy.
- Keep cows cool with misting systems or occasional hosing down.

4. Adjust Feeding:

- Provide feed during cooler parts of the day (early morning or late evening).
- Increase the proportion of high-energy feed to compensate for reduced intake.

Example:

In July, the temperatures in your region can soar to 100°F. You notice your cows are panting heavily and have reduced their milk output. You set up a misting system in their barn and start feeding them at dawn and dusk. You also add electrolyte supplements to their water. Within a few days, their condition improves, and milk production stabilizes.

Winter

Challenges:
- Cold stress
- Frozen water sources
- Increased feed requirements

Solutions:

1. Provide Warm Shelter:
   - Ensure barns and shelters are draft-free and have adequate bedding to keep animals warm.
   - Use heat lamps safely in extreme cold.

2. Prevent Water Freezing:
   - Use heated waterers or water heaters to keep drinking water from freezing.
   - Check water sources multiple times a day to ensure they remain unfrozen.

3. Increase Feed:
   - Increase the amount of feed to help animals generate more body heat.
   - Use high-energy feed to maintain body condition during cold weather.

4. Monitor Health:
- Watch for signs of frostbite, especially on ears and teats.
- Ensure animals are not shivering excessively, which indicates cold stress.

Example:
During a cold spell, the nighttime temperature drops to -10°F. You notice the water troughs are freezing over quickly. You install a heated water trough and add extra hay in the barn for bedding. You also increase the cows' feed ration by 20% to help them stay warm. The cows remain comfortable and healthy throughout the cold period.

Spring

Challenges:
- Mud and wet conditions
- Parasite outbreaks

Solutions:

1. Manage Mud:
- Create well-drained pathways and feeding areas.
- Use gravel or wood chips in high-traffic areas to reduce mud.

2. Parasite Control:
- Regularly deworm livestock according to a veterinarian's schedule.
- Rotate pastures to prevent the buildup of parasites in the soil.

3. Watch for Hoof Problems:
- Inspect hooves regularly for signs of rot or infection.
- Keep hooves trimmed to prevent issues caused by wet, muddy conditions.

Example:
Spring rains have turned your pasture into a muddy mess. Your cows are reluctant to leave the barn, and you notice a couple of them have developed hoof issues. You lay down gravel on the main pathways and feeding areas to reduce

mud. You also start a regular hoof inspection routine and consult your vet about a deworming schedule. This proactive approach keeps your cows' hooves healthy and minimizes the impact of the wet conditions.

Fall

Challenges:
- Transition from warm to cold
- Preparing for winter

Solutions:

1. Gradual Transition:
   - Start feeding additional hay as temperatures drop to help cows adjust.
   - Begin winterizing water sources and shelters before the first frost.

2. Health Checks:
   - Perform a thorough health check on all animals, looking for any signs of illness or parasites.

- Vaccinate against common diseases as recommended by your vet.

3. Stock Up:
- Ensure you have enough feed and bedding to last through the winter.
- Check and repair any shelter or fencing that might be damaged during winter storms.

Example:
As fall progresses, nighttime temperatures begin to drop. You start adding extra hay to the cows' diet and check that all waterers are working correctly. You also inspect and reinforce the barn and fencing. By the time winter arrives, your cows are well-prepared, and you have enough supplies to last through the season.

Practical Exercises

To help you become adept at managing seasonal variations, try these practical exercises:

1. Temperature Log:

- Keep a daily log of temperatures and observe how your cows respond to different conditions. Note any changes in behavior, milk production, or health issues.
- Use this log to anticipate problems and take preventative measures.

2. Water Monitoring:
- Monitor and log the water intake of your cows. Ensure that they always have access to clean, fresh water, adjusting for weather conditions.
- Implement and test a water heating system during the fall to prepare for winter.

3. Feed Adjustments:
- Experiment with different feeding schedules and types of feed. Note how changes affect milk production and cow health.
- Create a seasonal feeding plan based on your observations and the advice of a nutritionist or vet.

4. Shelter Assessment:

- Regularly inspect your shelters and barns for signs of wear and potential issues.

- Practice setting up temporary shade or windbreaks to ensure you're prepared for sudden weather changes.

Troubleshooting Common Problems

Even with the best preparation, problems can arise. Here are some common issues and how to troubleshoot them:

1. Heat Stress:
Symptoms: Panting, drooling, decreased milk production.
Solution: Provide immediate cooling with fans or misting. Ensure cows have access to shade and cool water. Increase air circulation in barns.

2. Frozen Water Sources:
Symptoms: Reduced water intake, dehydration signs.

Solution: Use heated waterers or regular checks to break ice. Insulate water lines and troughs.

3. Mud-Related Hoof Problems:
Symptoms: Lameness, swelling, foul odor from hooves.
Solution: Clean and dry hooves thoroughly. Apply hoof treatments as recommended by a vet. Improve drainage in muddy areas.

4. Parasite Infestations:
Symptoms: Weight loss, poor coat condition, diarrhea.
Solution: Implement a regular deworming schedule. Rotate pastures and maintain good hygiene practices.

By staying vigilant and proactive, you can effectively manage the challenges that come with seasonal variations and weather conditions in backyard dairy farming. Each season brings its unique set of obstacles, but with proper planning and care, your dairy farm can thrive year-round.

9.2 Reproduction and Breeding Challenges

Breeding your backyard dairy animals successfully is essential to maintaining a productive and healthy farm. Whether you're raising goats, cows, or sheep, understanding and overcoming the challenges associated with reproduction is crucial. This chapter will guide you through common problems, practical solutions, and hands-on exercises to ensure your breeding program is effective.

Understanding the Basics of Reproduction

Before diving into troubleshooting, it's important to understand the basic principles of animal reproduction:

1. Estrous Cycle: The reproductive cycle in female animals, which includes periods of fertility (estrus or "heat") and non-fertility. For *example*, cows typically have a 21-day cycle, goats around 21 days, and sheep about 17 days.

2. Gestation Period: The length of time an animal is pregnant. Cows have a gestation period of approximately 9 months, goats about 5 months, and sheep around 5 months.

3. Breeding Seasons: Some animals, like sheep and goats, are seasonal breeders, while cows can breed year-round.

Common Reproduction Challenges

1. Identifying Heat

Problem: Missing the signs of heat in your animals can result in missed breeding opportunities.

Solution:
- ***Observation:*** Spend time watching your animals. Signs of heat include restlessness, mounting behavior, mucus discharge, and swollen vulva.
- ***Heat Detection Aids:*** Use tools like heat detection patches, teaser animals (vasectomized

males), or hormone treatments prescribed by a veterinarian.

Exercise:
- Spend 15 minutes each morning and evening observing your animals during their suspected breeding season. Note down any signs of heat in a logbook.

2. Infertility Issues

Problem: Some animals may not conceive despite repeated breeding attempts.

Solution:
- ***Nutrition:*** Ensure animals receive a balanced diet with adequate vitamins and minerals. Deficiencies can impact fertility.
- ***Health Checks:*** Regular veterinary check-ups to identify any underlying health issues such as infections or hormonal imbalances.
- ***Breeding Records:*** Maintain detailed records of each breeding attempt, including dates, observed heats, and any veterinary treatments.

Exercise:
- Review your animals' diet and consult with a vet to create a tailored nutrition plan. Implement any necessary changes and monitor for improvements in fertility.

3. Pregnancy Detection

Problem: Not knowing if an animal is pregnant can lead to inefficient farm management.

Solution:
- *Manual Palpation:* Learn to palpate your animals to feel for pregnancy signs. This method requires experience and gentle handling.
- *Ultrasound:* Schedule ultrasounds with a veterinarian for early and accurate pregnancy detection.
- *Behavioral Changes:* Notice changes in behavior and physical appearance, such as decreased appetite and udder development.

Exercise:

- Work with your vet to practice manual palpation techniques on a few animals. Over time, you'll become more adept at detecting pregnancies on your own.

Advanced Troubleshooting

1. Handling Difficult Births

Problem: Complications during labor can endanger both the mother and offspring.

Solution:
- ***Preparation:*** Have a birthing kit ready, including gloves, lubricant, clean towels, and antiseptics.
- ***Know When to Intervene:*** Learn the signs of a normal birth versus when there is a problem (e.g., prolonged labor, abnormal positioning).
- ***Veterinary Assistance:*** Have your vet's contact information readily available for emergencies.

Exercise:

- Create a birthing kit and familiarize yourself with its contents. Watch videos or attend workshops on assisting with difficult births.

2. Managing Postpartum Health

Problem: Postpartum complications can affect milk production and overall health.

Solution:
- *Monitoring:* Keep a close eye on the mother for signs of distress, infection, or mastitis (udder infection).
- *Nutrition:* Provide high-quality feed and supplements to support recovery and lactation.
- *Clean Environment:* Ensure the birthing area is clean and dry to prevent infections.

Exercise:
- Develop a postpartum care plan that includes daily health checks, a nutritional regimen, and a clean living environment for the new mother and her offspring.

Practical Steps for a Successful Breeding Program

1. Set Clear Goals: Define your breeding goals, whether it's to increase milk production, improve herd genetics, or expand your farm.
2. Choose Quality Breeding Stock: Select animals with good health, temperament, and proven reproductive success.
3. Maintain Detailed Records: Track breeding dates, heat cycles, pregnancy confirmations, and birth outcomes.
4. Implement Biosecurity Measures: Prevent the introduction of diseases by quarantining new animals and practicing good hygiene.
5. Educate Yourself: Stay informed about the latest breeding techniques and veterinary practices through books, online courses, and local farming groups.

Reproduction and breeding challenges are a natural part of backyard dairy farming. By understanding the basics, identifying common problems, and implementing practical solutions, you can enhance the reproductive success of your herd. Remember, patience and continuous learning are key to becoming an expert in backyard dairy farming. Keep detailed records, work closely with your veterinarian, and never hesitate to seek help when needed. Happy farming!

9.3 Addressing Milking and Production Issues

This section is dedicated to addressing milking and production issues that may arise in your backyard dairy operation. We'll walk you through common problems, provide practical steps for troubleshooting, and suggest exercises to help you become proficient in managing your dairy animals. Whether you're a novice or an experienced farmer, this guide aims to equip you with the knowledge and skills needed to maintain a healthy and productive dairy farm.

Common Milking and Production Issues

1. Low Milk Production

Symptoms: Reduced milk yield from your dairy animals.

Possible Causes:
- ***Nutritional Deficiencies:*** Insufficient feed quality or quantity.
- ***Health Problems:*** Mastitis, metabolic disorders, or other illnesses.
- ***Poor Milking Practices:*** Inconsistent milking times, improper technique.
- ***Environmental Stress:*** Extreme weather, inadequate shelter.

Troubleshooting Steps

1. Assess Diet: Ensure your animals receive a balanced diet rich in energy, protein, vitamins, and minerals. Consult a livestock nutritionist if necessary.

2. Health Check: Regularly monitor your animals for signs of illness. Conduct routine health checks and consult a veterinarian for any concerns.

3. Milking Routine: Establish a consistent milking schedule. Train yourself and any helpers on proper milking techniques to ensure complete milk let-down.

4. Comfort and Environment: Provide adequate shelter and a comfortable living environment. Ensure access to clean water and shade during hot weather.

Exercise: Create a weekly feeding plan for your dairy animals, including different feed types and quantities. Track milk production daily to identify any correlations between diet and yield.

2. Mastitis

Symptoms: Swollen, hot, and painful udder, abnormal milk (clots, blood, or pus), reduced milk yield.

Possible Causes:
- *Bacterial Infection:* Commonly caused by poor hygiene during milking.
- *Injury:* Trauma to the udder can lead to infection.

- *Improper Milking:* Incomplete milking or using dirty equipment.

Troubleshooting Steps:
1. Hygiene: Ensure hands, udder, and milking equipment are thoroughly cleaned before and after milking. Use disinfectants approved for dairy use.
2. Proper Milking Technique: Milk gently and completely to avoid injury and ensure thorough milk extraction.
3. Treat Infected Animals: Isolate the affected animal and consult a veterinarian for appropriate treatment, which may include antibiotics.
4. Preventive Measures: Regularly check udders for early signs of infection. Implement a mastitis control program, including proper teat dipping and dry cow therapy.

Exercise: Practice the proper milking technique on a dummy udder or with a seasoned farmer. Simulate the steps of cleaning, milking, and post-milking care to build muscle memory.

3. Inconsistent Milk Quality

Symptoms: Milk with off-flavors, abnormal consistency, or irregular fat content.

Possible Causes:
- ***Feed Contamination:*** Certain feeds can alter the taste of milk.
- ***Poor Hygiene:*** Contaminated milking equipment or environment.
- ***Health Issues:*** Diseases affecting milk composition.

Troubleshooting Steps:
*1. **Feed Quality:*** Avoid feeding animals strong-smelling plants like onions or garlic. Ensure feed storage is dry and clean to prevent mold and spoilage.
*2. **Sanitation:*** Sterilize milking equipment regularly. Ensure the milking area is clean and free from contaminants.
*3. **Health Monitoring:*** Regularly check animals for signs of diseases that could affect milk

quality. Conduct milk tests to monitor somatic cell counts and other indicators.

Exercise: Conduct a mock inspection of your milking area and equipment. Identify potential contamination sources and create a cleaning schedule to maintain hygiene standards.

4. Difficulty in Milking

Symptoms: Animals reluctant to be milked, kicking, or incomplete milk let-down.

Possible Causes:
- *Stress:* Environmental stressors or handling issues.
- *Injury or Illness:* Pain from injury or underlying health issues.
- *Improper Handling:* Rough or inconsistent milking practices.

Troubleshooting Steps:

1. Calm Environment: Create a quiet and calm milking environment. Handle animals gently and speak softly.

2. Health Check: Check for injuries or illnesses that may cause discomfort during milking.

3. Milking Routine: Train animals to a consistent milking routine. Use positive reinforcement to reward good behavior.

4. Handling Techniques: Learn proper animal handling techniques to reduce stress and encourage cooperation during milking.

Exercise: Spend time handling your dairy animals outside of milking times. Practice gentle handling and reward calm behavior to build trust and reduce stress.

Practical Problem-Solving Strategies

Regular Monitoring and Record Keeping
Keeping detailed records of your dairy operation is essential for identifying and addressing issues promptly. ***Track the following:***
- ***Milk Production:*** Daily yield per animal.

- *Health Status:* Any signs of illness or injury.
- *Feeding Records:* Types and amounts of feed given.
- *Milking Schedule:* Times and techniques used.

Exercise: Create a logbook or use a digital tool to record daily observations. Review these records weekly to spot trends and address issues proactively.

Regular Health Checks

Implement a routine health check protocol for your dairy animals. This includes:
- *Physical Examination:* Check for signs of illness, injury, or abnormal behavior.
- *Udder Inspection:* Look for signs of mastitis or other udder issues.
- *Fecal Monitoring:* Regularly check for parasites or digestive issues.

Exercise: Develop a checklist for health checks and conduct them weekly. Familiarize yourself with normal and abnormal signs to improve your diagnostic skills.

Consulting Experts

Don't hesitate to seek help from experts such as veterinarians, livestock nutritionists, and experienced dairy farmers. They can provide valuable insights and support for more complex issues.

Exercise: Build a network of contacts including local vets, agricultural extension services, and fellow dairy farmers. Attend workshops and join online forums to stay informed about best practices and new developments in dairy farming.

Continuing Education

Stay updated with the latest knowledge and techniques in dairy farming. Read books, attend seminars, and participate in training programs.

Exercise: Allocate time each month for self-education. Set learning goals and track your progress. Share your knowledge with others to reinforce your learning and contribute to the community.

Milking and production issues are a natural part of dairy farming, but with the right knowledge and approach, they can be effectively managed. By implementing the troubleshooting steps, exercises, and practical strategies outlined in this guide, you'll be well-equipped to address these challenges and ensure a productive and healthy dairy operation. Remember, patience and continuous learning are key to becoming an expert in backyard dairy farming. Happy farming!

Chapter 9

Scaling Up and Expanding

10. Taking Your Backyard Dairy Farming to the Next Level

As a backyard dairy farmer, you may have started with just a couple of cows or goats, providing fresh milk for your family. Now, you're thinking about scaling up and expanding your operations. This comprehensive guide will walk you through the steps to take your dairy farm to the next level, from expanding your herd to exploring commercial opportunities.

1. Expanding Your Herd and Production Capacity

a. Assessing Your Current Setup

- **Space:** Ensure you have enough land to accommodate more animals. Each cow needs about 1.5 to 2 acres for grazing, while goats require less, around 0.5 acres per goat.

- *Shelter:* Expand your barns or shelters to provide adequate space. Overcrowding can lead to stress and health problems.
- *Feed and Water:* Plan for increased feed and water needs. A lactating cow can drink up to 50 gallons of water a day and will need a balanced diet to maintain milk production.

b. Selecting the Right Animals

- *Breeds:* Choose breeds known for high milk production. For cows, Holsteins and Jerseys are popular. For goats, consider Saanens or Alpines.
- *Health:* Purchase animals from reputable breeders. Ensure they are healthy and have been tested for common diseases like bovine tuberculosis or brucellosis.

c. Practical Steps for Expansion

1. Budgeting: Calculate the costs for additional animals, expanded facilities, feed, and healthcare.

2. Quarantine New Animals: Keep new additions separate for at least two weeks to monitor for any signs of illness.

3. Gradual Integration: Introduce new animals gradually to avoid stress and fighting among your herd.

Example: If you currently have two cows and want to expand to six, start by securing an additional 8 acres of land. Build additional shelter to house four more cows and stock up on feed and supplements. Budget for the purchase price of four cows, plus the extra feed and healthcare costs.

2. Exploring Commercial Opportunities

a. Selling Milk and Dairy Products
- ***Regulations:*** Research local regulations regarding the sale of raw milk. Some areas require pasteurization or special permits.
- ***Products:*** Consider diversifying into cheese, yogurt, butter, or ice cream. These value-added products can increase your profit margins.
- ***Marketing:*** Develop a brand for your dairy products. Use social media, local markets, and word of mouth to attract customers.

b. Setting Up a Dairy Processing Area

- *Equipment:* Invest in pasteurizers, cream separators, cheese presses, and storage facilities.
- *Hygiene:* Maintain strict cleanliness standards to ensure the safety and quality of your products.
- **Packaging:** Choose attractive and practical packaging to appeal to consumers and meet regulatory standards.

Example: If you decide to produce cheese, start by learning the basics through online courses or local workshops. Invest in a cheese press and appropriate molds. Begin with simple cheeses like feta or mozzarella, which are easier to make and popular with consumers.

3. Collaborating with Local Dairy Networks

a. Joining Dairy Associations and Cooperatives

- *Benefits:* Gain access to shared resources, bulk buying discounts, and marketing support.

- *Networking:* Connect with other dairy farmers to share knowledge, experiences, and best practices.
- *Training:* Take advantage of workshops and training sessions offered by these groups to improve your skills.

b. Participating in Local Markets and Events

- *Farmers' Markets:* Sell your products directly to consumers and get valuable feedback.
- *Community Events:* Sponsor or participate in local fairs, festivals, or food events to increase your visibility.

c. Establishing Partnerships

- *Local Stores:* Partner with local grocery stores, health food stores, and restaurants to stock your products.
- *CSA Programs:* Collaborate with Community Supported Agriculture (CSA) programs to provide regular dairy product deliveries to subscribers.

Example: Join your local dairy farmers' association to benefit from bulk purchasing of feed and supplies. Participate in a weekly farmers' market to sell your milk and cheese directly to the community, building a loyal customer base.

Practical Exercises to Boost Your Dairy Farming Skills

1. Record-Keeping: Start a detailed log of your herd's health, milk production, feeding schedules, and expenses. Use this data to identify trends and make informed decisions.

2. Farm Visit: Visit a larger dairy farm to observe their operations and gain insights into scaling up your own farm.

3. Workshops and Webinars: Regularly attend dairy farming workshops and webinars to stay updated on the latest techniques and technologies.

Example Exercise: Keep a daily log for a month, noting each cow's milk production, any health issues, and feed consumption. Analyze the data to determine if any changes in diet or care practices correlate with changes in milk yield.

By following these steps and incorporating these practical exercises, you can successfully scale up your backyard dairy farming operation. Whether you're expanding your herd, exploring new commercial opportunities, or collaborating with local networks, there's a world of potential waiting for you in the dairy industry.

www.ingramcontent.com/pod-product-compliance
Lightning Source LLC
Chambersburg PA
CBHW050056230526
45470CB00004B/1554